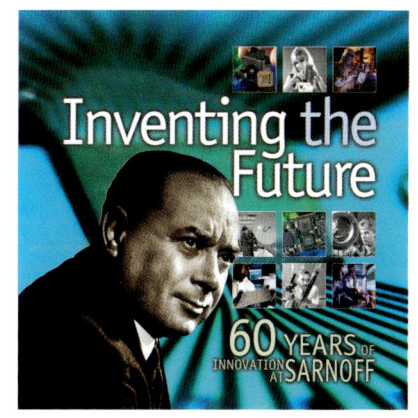

Inventing the Future

60 Years of Innovation at SARNOFF

Editor
Thomas V. Lento, PhD

Preface by
Dr. Henry Kressel

Introduction by
Satyam C. Cherukuri, PhD

Historical Consultant
Alexander B. Magoun, PhD

© 2006 Sarnoff Corporation

Published in the United States by
Sarnoff Corporation, Princeton, New Jersey

Unauthorized reproduction of this book is forbidden by law
without written permission from the publisher.

Library of Congress Control Number: 2006924751

ISBN 0-9785463-0-X

INVENTING THE FUTURE

"There is no security in the future in the mere knowledge of today. There is hope and opportunity in what we can learn tomorrow."

David Sarnoff
(1941)

INVENTING THE FUTURE

Acknowledgements

Sarnoff Corporation thanks and acknowledges the many people who contributed to the breadth and depth of information capsulated in this book. Through their dedication and knowledge, we have been able to preserve the rich history of innovation and share the contributions Sarnoff has made to the world.

Al Acampora
Roger Alig
Leslie Avery
Robert Bartolini
Peter Burt
Raymond Camisa
Curt Carlson
James Carnes
Charles Carroll
Carmen Catanese
George Cody
Deborah Dinardo
Douglas Dixon
Joseph Dresner
Vince Endres
Ronald Enstrom
Michael Ettenberg
Robert Harris

Bernie Hershenov
Lori Hewitt
Jim Kaba
Grzegorz Kaganowicz
Iz Kalish
Bernard Lechner
Thomas Lento
Aaron Levine
Peter Levine
Steven Lipp
Gary Looney
Charles Magee
Alexander Magoun
Herb Maruska
James Matey
Dennis Matthies
Jock McFarlane
James Oerth

Norman Oldfield
Steven Perna
Harry Pinch
Jeremy Pollack
John van Raalte
Glenn Reitmeier
John Riganati
Ronald Roach
Anthony Robbi
Dorothy Sadley
Robert Simms
David Southgate
Fred Vannozzi
William Webster
Richard Williams
Peter Wojtowicz
Niel Yocom

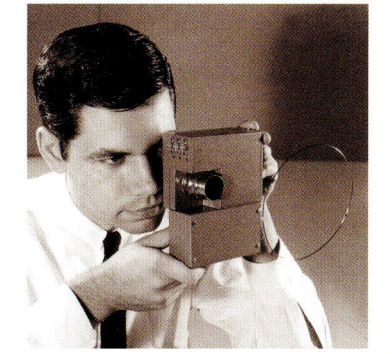

Contents

The Video Revolution Sarnoff Invents an Industry	**10**
Microelectronics Revolution Transistors, ICs, CMOS Process	**18**
Information On A Beam of Light Lasers and LEDs	**28**

Preface
Productive Science
6

Introduction
Innovation Goes Global
8

Above: William Homa operates the world's first solid-state camera in 1966.

Teaching Computers to See Computer Vision Technology	**40**
The World On Display Picture Tubes, Flat Panels, and Projectors	**48**
The Bionic Future Diagnoses, Drugs, Devices	**56**
Extending the Power of Sight Electronic Cameras and Imagers	**64**
Transmitting Intelligence Communications and Computing	**74**
Elements of Innovation Electronic Materials and Processes	**84**
Stay Tuned The Future Is Still Ahead	**95**

INVENTING THE FUTURE

Productive Science

Preface
Dr. Henry Kressel
Managing Director,
Warburg Pincus

Dr. Henry Kressel is a managing director at Warburg, Pincus & Co, LLC, a diversified venture capital firm. Prior to 1983 he was vice president at RCA Laboratories, responsible for worldwide R&D in optoelectronics, power devices, integrated circuits and associated software. Dr. Kressel holds 31 U.S. patents for electronic and optoelectronic devices, and pioneered the first practical semiconductor lasers. He was the founding president of the IEEE Laser and Electro-Optics Society (LEOS) and is a member of the National Academy of Engineering.

The achievements described in this book are testimony to the productivity of a unique organization, one that has mastered the art of creating industrial innovations and managing their commercial realization.

RCA Laboratories, the predecessor of Sarnoff Corporation, opened its doors in 1942. Its formation reflected David Sarnoff's total focus on building RCA around innovative electronic technologies.

The results surpassed even his expectations. During its first 45 years, when it was RCA's corporate research center, the Labs pioneered many of the innovations that comprise the foundation of our modern world infrastructure. As an independent company, Sarnoff Corporation has continued this tradition of major innovations on behalf of government and industrial partners.

The organization's success is based on a few basic operating principles that have guided its management since the beginning. These include market focus, product involvement, breadth of expertise, strategic vision, and rapid assimilation of new developments.

Market focus is the most important of these. History has shown that industrial research laboratories that are divorced from market reality eventually fail. The management of the Laboratories was acutely aware of this pitfall.

They avoided it by creating a unique culture and management style that successfully matched the individual's freedom to innovate with the strategic direction of the company. While the scientific specialists in various disciplines at the Labs were among the best in the world, their contributions went beyond research and discovery. They were also expected to help bring concepts from the laboratory to the market.

It was not unusual to see scientists working on the floor of RCA factories to help in a product launch. I remember working the night shift to train factory personnel during the introduction of the early silicon transistors, and later during production of semiconductor lasers.

Management also understood that excessively narrow specialization is the enemy of creativity. An important factor in the success of the Laboratories was its ability to develop talent over time to meet changing technological and commercial requirements. It was not unusual for scientists to change fields every few years to develop new areas.

This had an important advantage: the experience gained in one field could be transferred to a new area, thus providing new insights. In my case, for example, over the years I had the opportunity to move my research work from transistors to microwave devices, then to semiconductor lasers and light emitting diodes, from there to solar cells, and finally to integrated circuits. At each stage I was able to apply lessons from one field to problems in another. The experience with semiconductor materials and processes helped me find ways to produce commercially viable laser diodes.

Scientists were also involved in product strategy. Management frequently initiated programs to build applications so that researchers could demonstrate the commercial potential of their innovations. For example, in the early 1970s we created prototype fiber optical systems to show how our newly developed RCA commercial semiconductor lasers might be used in communications.

This culture of flexibility was the direct result of having managers with broad knowledge of both technologies and markets. It was expected that senior technical staff members would be fully cognizant of any developments in their fields that might impact the company's business, and be ready to launch programs to address new market opportunities.

Thanks to its flexible management structure, the Labs could react very quickly to external developments. For example, the development of digital television was launched at the Laboratories in the late 1970s, years ahead of most competing efforts.

Now, in this book, this record can finally speak for itself. The hundreds of breakthrough technologies and products listed in these pages show how creative and productive the organization has been over the past 60-plus years.

Yet even this is only part of the story. As the central corporate research organization for RCA, the Labs had amassed a portfolio of over 20,000 patents when the General Electric Company acquired RCA in 1985. It is interesting to note that this intellectual property brought GE annual licensing revenues in excess of $250 million per year.

You could not ask for more convincing proof of the success of RCA Laboratories and Sarnoff Corporation in combining good science with commercial savvy. That is the best kind of innovation.

Innovation Goes Global

Introduction
Satyam C. Cherukuri, PhD
President and CEO
Sarnoff Corporation

Dr. Satyam Cherukuri joined Sarnoff as a researcher in 1989. In 1998 he was named Managing Director of the company's Life Sciences and Systems unit, which he was instrumental in developing. He became Sarnoff's Chief Operating Officer in 2001, and President and CEO in 2002. Dr. Cherukuri is Chairman of the Executive Council of the Conference Board. He also sits on the Board of Directors for Ness Technologies, Inc. (NASDAQ: NTSC), a global provider of IT services and solutions.

It is now possible to source new ideas, new technologies, and new products worldwide. The technology supply chain has been realigning, making it critical for companies to be connected into the global innovation network. Those that don't engage now are doomed to watch helplessly as the competition passes them by.

Major corporations are struggling to adapt before it's too late. They are looking for ways to reap the benefits of this commercial revolution rather than fall behind in the race to develop successful new products.

Investing in the Future
Sarnoff has developed a highly effective approach to innovation on a global scale. We've established subsidiaries and affiliates in Europe and Asia to build a global innovation supply chain with human capital worldwide.

Through this international network we can act as a conduit between our clients and a pool of innovators, partners, and potential customers around the world.

Sarnoff has continually adapted to changing market conditions. Our breakthrough technologies and products were usually created with international markets in mind. This book summarizes these achievements.

R&D Decentralized
The corporate R&D model, invented by Edison over 100 years ago and embodied by Sarnoff and other renowned laboratories, was responsible for generating fundamental breakthroughs in electronics,

communications, computers, pharmaceuticals, materials — in fact, every industry.

For 45 years after its founding in 1942, Sarnoff's innovations helped RCA, our parent company, revolutionize the electronics and communications industries. The outcomes included color TV and CRTs, lasers, color LCDs, the CMOS process, silicon solar cells, and computer vision.

In the 1970s, companies met the increased demand for innovation by looking beyond corporate R&D. They licensed new technology, or bought small, entrepreneurial companies with promising products.

In the 1990s venture capital-funded entrepreneurial companies drove the stock market to dizzying heights while they created new products and services for a seemingly insatiable market.

A new model, distributed innovation, has emerged with Sarnoff as an early pioneer. Having spun out from RCA in 1987, we became a contract research provider, developing technology for government and commercial clients in such areas as satellite broadcasting, computational drug design, and HDTV. We also founded over 20 start-up companies, often in partnership with our clients, to bring innovations to market.

The Global Opportunity

It all came to an abrupt and sobering halt in 2001 with the collapse of the Internet bubble. The economic crisis accelerated a growing trend toward internationalization. The flood of foreign-born scientists and entrepreneurs to U.S. centers of innovation began to ebb. The worldwide communications infrastructure that came out of Silicon Valley and other U.S. innovation centers made it possible for innovators to stay home, yet stay in touch. It helped create a global ecosystem for creating and consuming technology. The internationalization of technical skills and business operations has helped eliminate the cost of production facilities as a barrier to innovation.

The story has now come full circle. As this book shows, Sarnoff Corporation has long been a vital international force for innovation, working at the forefront of technology and its commercial application for over 60 years.

We have operated in, even pioneered, each of the three innovation modes: corporate R&D, distributed services, and global collaboration. This first-hand experience is a tremendous asset as we help clients face the challenge of transitioning to a global innovation model.

We look forward to putting our international network of resources at the service of our partners to create tomorrow's successful products.

The Video Revolution

SARNOFF INVENTS AN INDUSTRY

1999 Emmy for compliance bitstreams. *Above*: David Sarnoff and his Emmy for contributions to TV broadcasting.

"Now we add sight to sound."

The year was 1939, the occasion was the New York World's Fair, and the event was the announcement of RCA's all-electronic television system. The speaker? David Sarnoff, RCA's president and the namesake of Sarnoff Corporation.

His words were the precise distillation of a much larger vision. Television would totally change how people perceived the world. Even though no one could have foreseen TV's full impact at its introduction, David Sarnoff understood its significance.

Today hardly any aspect of life is untouched by television. Wars have been fought on TV, elections contested, the values and lifestyles of societies revealed and recorded, new cultural artifacts created. Whole industries have sprung up around it. Its core technology, electronic displays, is at the heart of the computer revolution.

From the very beginning Sarnoff Corporation has been a major source of video technology. Nine Emmy® awards for technical excellence bear witness to its efforts on behalf of television. Its video innovations for non-TV use are equally pervasive. Sarnoff video display technology sits on literally every computer desk. Its vision technology helps spot intruders, guide vehicles, and improve medical imaging.

SIGNIFICANT SARNOFF VIDEO INNOVATIONS

1944 Video camera tube for guided missiles, later used in TV studios.
1947 Proposal for an electronic, monochrome-compatible color television system, the basis of the NTSC standard.
1950 Shadow mask CRT, the first color television picture tube.
1953 FCC approval of the NTSC color television standard; Sarnoff created the fundamental technology and design of the color-encoding system, cameras, CRT (including phosphors), receivers.
1968 LCD technology, including basic chemistry and first commercial applications for today's dominant flat-panel display.
1974 CCD image sensor, adapting an existing technology to commercial use.
1981 RCA introduces CED VideoDisc product into consumer market.
1984 Pyramid-based image processing, the basis for vision technology and real-time processing of video.
1986 Stereo sound for broadcast television.
1991 Advanced Digital HDTV, a predecessor of the ATSC digital TV standard approved by the FCC five years later.
1992 Digital direct broadcast satellite TV, at that time the fastest growing consumer electronics product in history.
1993 Real-time insertion of images into live video.
1996 FCC approves ATSC system as the standard for U.S. digital television. Sarnoff was a leader of the Grand Alliance, a seven-member consortium that developed the new system.
1997 VideoBrush, commercial software for PCs to turn live video into panoramic still images for display and printing.
1999 Integrated circuit chipset designed in collaboration with Motorola to receive digital TV signals and convert the images for display on analog TVs.
2000 Acadia® I, world's fastest vision processing chip for real-time video processing in desktop PCs and other consumer products.
2003 Watermarking for digital cinema that survives compression, encoding, and Internet distribution.

GLOBAL IMPACT

Television has become a force in the world economy. A major consumer electronics manufacturing industry arose around it. In 1948 there were 350,000 TV sets in American homes; in 2000 there were 100 million. There is one set for every four people in the world, or nearly 1.8 billion televisions.

It has generated new growth in related businesses. The most obvious example is advertising. In 1941, in the first commercial broadcast in the U.S., Sarnoff's NBC-TV ran a 10-second commercial for a Bulova watch and netted $7. Sixty years later advertisers paid $24 billion for spots on TV.

As with any successful new system, television stimulated technology spinoffs such as video displays, video cameras, video recorders, recordable video media, and cable and satellite delivery systems. Many of these found important applications outside of television, such as medical diagnostics, security and surveillance, and scientific and technical imaging.

RELATED TOPICS

CRT, LCD, Flexible Plastic Displays: see *The World on Display*
CCD Image Sensor: see *Extending the Power of Sight*; also see *Information on a Beam of Light*
Pyramid Processing, VideoBrush, Acadia I: see *Teaching Computers to See*
Digital TV chipset: see *The Microelectronics Miracle*

ABOVE: Demonstration of TV weather channel, 1958, combining Labs expertise in television and airborne radar systems.

INVENTING THE FUTURE

THE VIDEO REVOLUTION

TOP LEFT: On-screen menu for video-cassette recording, 1982. The helical-scan head used in all VCRs was developed at Sarnoff.

TOP RIGHT: Paul Weimer demonstrates his vidicon camera in 1950. The vidicon, a simple, inexpensive, and effective imager, was used in industrial, military, and closed-circuit video systems into the 1980s.

RIGHT: Paul Mitnaul demonstrates RCA's VideoDisc, a predecessor of the DVD, in 1980.

BELOW: Harry Olson shows David Sarnoff the world's first videotape player for the home, developed as a 65th "birthday present" to Sarnoff in 1956.

LEFT: Setting up a microwave antenna at Princeton University's Graduate School in 1946 to receive an all-electronic color television signal from the Labs, two miles away.

INVENTING THE FUTURE

LEFT: Sarnoff was a leader of the seven-member Grand Alliance that developed the HDTV standard in 1995.

BELOW LEFT: Sarnoff digital video interactive (DVI) technology was simple enough for a child to use (1986).

BELOW RIGHT: The DVI team that pioneered the first interactive CDs. Sarnoff spun off the DVI technology to Intel in 1988.

RIGHT: Sarnoff Compliance Bitstreams are the de facto U.S. standard for verifying that receivers will reliably decode and display digital TV and HDTV signals.

THE VIDEO REVOLUTION

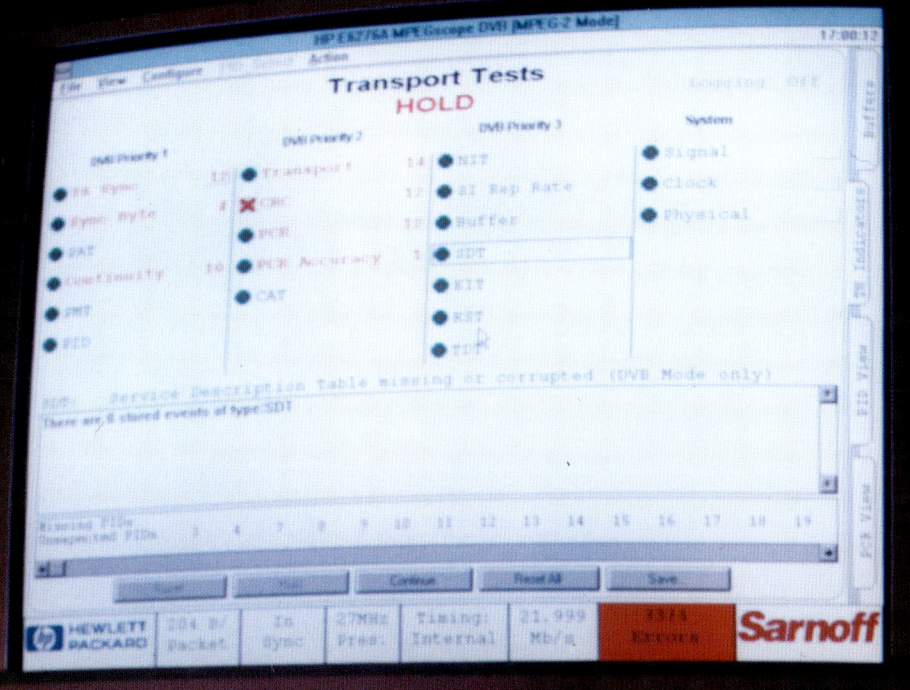

INVENTING THE FUTURE

Sarnoff was instrumental in developing direct-to-the-home digital satellite TV, the fastest growing consumer product in history before the DVD.

THE VIDEO REVOLUTION

ABOVE LEFT: The technical leaders of the Grand Alliance pose in Sarnoff's field lab: standing L-R: Wayne Luplow (Zenith), Glenn Reitmeier (Sarnoff), Bob Rast (General Instruments), Terry Smith (Sarnoff); seated L-R: Ralph Cerbone (AT&T), Jae Lim (MIT), Aldo Cugnini (Philips).

ABOVE: Al Acampora with Sarnoff's priority processor board for MPEG picture information, part of the advanced digital high-definition TV (ADHDTV) system, a forerunner of the Grand Alliance digital TV system that became the official U.S. standard.

Frank Lang adjusts an antenna during DTV tests in Washington, D.C.

LEFT: JNDmetrix™ technology predicts how humans will react to digitally processed video or other images, allowing program originators to adjust the processing for highest perceived quality.

INVENTING THE FUTURE

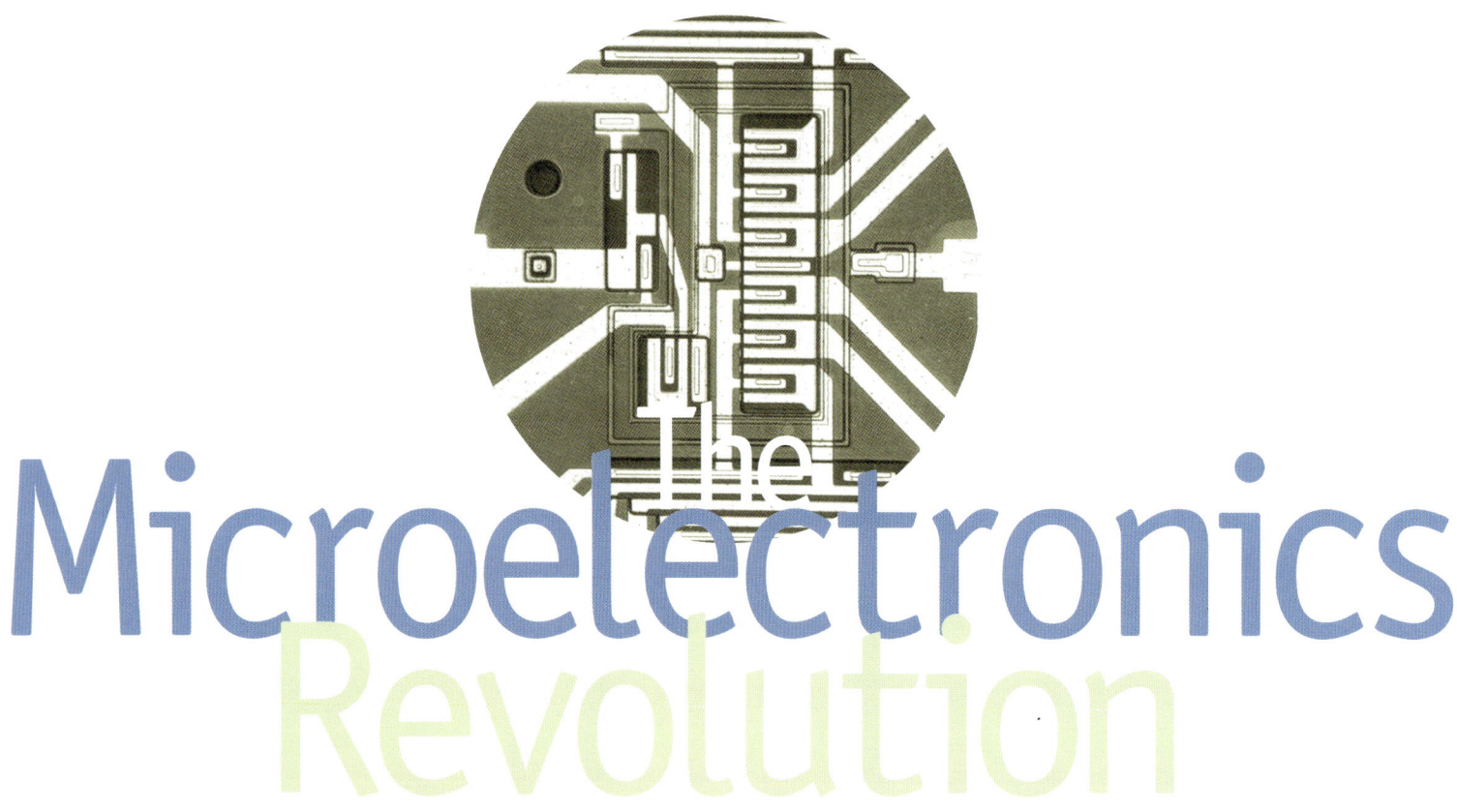

The Microelectronics Revolution

TRANSISTORS, ICs, CMOS PROCESS

Power is no longer synonymous with size.

In the early 1950s you could buy a "portable" radio, but it was bulky, required frequent battery changes, and its vacuum tubes let it double as a foot-warmer. Five years later the world was carrying a transistorized radio in its shirt pocket.

In the 1960s a computer filled a climate-conditioned room and was limited to basic number-crunching. Today people carry computers in their briefcases that can create graphics, encode video, and communicate over the Internet, all almost instantly.

The astonishing evolution of electronic devices during the last 40 years has transformed the way people live, work, play, communicate, and relate to their world. It all grew out of the microelectronics miracle, driven in large part by Sarnoff innovations.

In the early days, working with RCA's Semiconductor Materials Division, Sarnoff developed better materials and processes for transistors and diodes, including defect-free crystals and better doping and cleaning methods. That's when radios and TVs abandoned vacuum tubes in favor of smaller, cooler, more reliable solid-state devices.

But the biggest breakthrough came in 1964: the development of the standard complementary metal oxide semiconductor (CMOS) process at Sarnoff. CMOS ultimately gave manufacturers a way to create low-power digital "computers on a chip," or integrated circuits (ICs) containing millions of transistors. Today CMOS is used to make over 90% of the world's microprocessors and memory chips.

Building on its historical leadership in IC design, processes, and materials, Sarnoff continues to advance the state of the art in semiconductor technology. Its silicon intellectual property (IP) adds functionality to countless ICs; TakeCharge® technology puts more chips on each wafer at many of the world's leading foundries. Sarnoff also designs new chips for IC makers.

SIGNIFICANT SARNOFF MICROELECTRONICS INNOVATIONS

1952 Demonstration of high-frequency transistors for use in radios, television, and other consumer electronics.

1959 Development of tunnel diodes for use in computers and communication equipment.

1962 Thin-film transistor (TFT), widely used as drivers for liquid crystal displays (LCDs).

1962 Metal oxide semiconductor field effect transistor (MOSFET), the basis of C (complementary) MOS process.

1964 Development of the CMOS process, which later became the technology used to manufacture the vast majority of the world's ICs and DRAM chips.

1966 Invention of vapor-phase epitaxy, a standard technique for semiconductor fabrication.

1970 Development of Standard Clean 1 and Standard Clean 2 to remove metallic contaminants during semiconductor oxide growth.

1976 Charge-coupled device (CCD) adapted to commercial use.

1977 Silicon-on-sapphire (SOS) ICs, used extensively in space applications circuits due to their inherent resistance to radiation.

1982 Gallium arsenide (GaAs) microwave devices, the first solid-state power amplifiers (SSPAs) to be used in communications satellites.

1986 GEM technology to create form, fit, and function replacements of no-longer-available ICs in Department of Defense electronic systems with current Bi-CMOS techniques, without having to redesign entire systems.

1992 Ultra-low-power A/D converter, using deep submicron CMOS technology to achieve a hundredfold reduction in power requirements.

1997 CMOS Active Pixel Sensor (APS) imager, a "camera-on-a-chip" with outstanding dynamic range.

1999 Digital TV demodulator and decoder chips, designed with Motorola, to adapt analog TVs for the new digital and HDTV standard.

2000 TakeCharge® technology for IC design, substantially reduces manufacturing costs for ICs by shrinking dies, eliminating reworks, and fitting more chips per wafer.

2003 Prototype of uncooled CMOS infrared (IR) sensor based on MEMS technology demonstrated.

2004 Silicon IP for MPEG-4 based video on cell phones, other devices and for ATSC digital TV demodulation and decoding offered for licensing.

GLOBAL IMPACT

Sales of semiconductors topped $215 billion worldwide in 2004. The business is broadly international, with the 10 largest firms headquartered across the U.S., Europe, and Asia.

At the finished product level, manufacturer-to-dealer sales of consumer electronics products in the U.S. alone exceeded $125 billion for 2004.

But even economic facts as impressive as these only begin to suggest the titanic transformation of our world by the electronics revolution.

It gave new life to existing technology. Electronics gave the auto industry the engine controls it needed to meet tougher emissions standards in the 1970s, and opened the way for many other safety and convenience features. Everything from wristwatches to harbor dredges have gained new precision and performance from the replacement of mechanical controls with electronics.

The electronics revolution also gave birth to new industries. The manufacture of the components is just the starting point. Personal computers, cell phones, and numberless consumer products all owe their existence to the ICs that pack 100 million devices in a volume of less than 0.004 cubic inches.

As the circuits within these tiny devices get smaller, and their power and speed increase, Sarnoff Corporation continues to play a major role in their development and application.

RELATED TOPICS

Electronic materials: see *Elements of Innovation*
Television and video: see *The Video Revolution*
Optoelectronics: see *Information on a Beam of Light*
Cameras: see *Extending the Power of Sight*

ABOVE: Gerry Miller probes the connections on a new integrated circuit, c1995.

LEFT: c1967 close-up of RCA's CMOS integrated circuit, the first in the market when released in 1968.

INVENTING THE FUTURE

A technician probes the first thin-film transistors in the world. Invented as camera sensors by Sarnoff's Paul Weimer in 1961, TFTs are used today to control liquid crystals in LCD panels.

THE MICROELECTRONICS REVOLUTION

ABOVE: Breadboard model of RCA's CMOS 4, a 4-bit/word memory circuit, in 1966. Based on this work RCA's Solid-State Design Center in Somerville, NJ designed and made the first CMOS IC in 1968 and the first CMOS microprocessor in 1976.

LEFT: Greg Zak demonstrates a new photoreduction camera for the preparation of IC masks at Sarnoff in 1978.

BELOW LEFT: Raymond Dean explains the Labs' gallium-arsenide (GaAs) microwave amplifier in 1967. Based on the Gunn Effect, it was the first solid-state amplifier to reliably generate 10 to 40-gigahertz. It was used for high-definition radar and satellite communications.

BELOW RIGHT: GaAs field-effect transistor, c1976. RCA's Fred Sterzer pioneered this efficient microwave device.

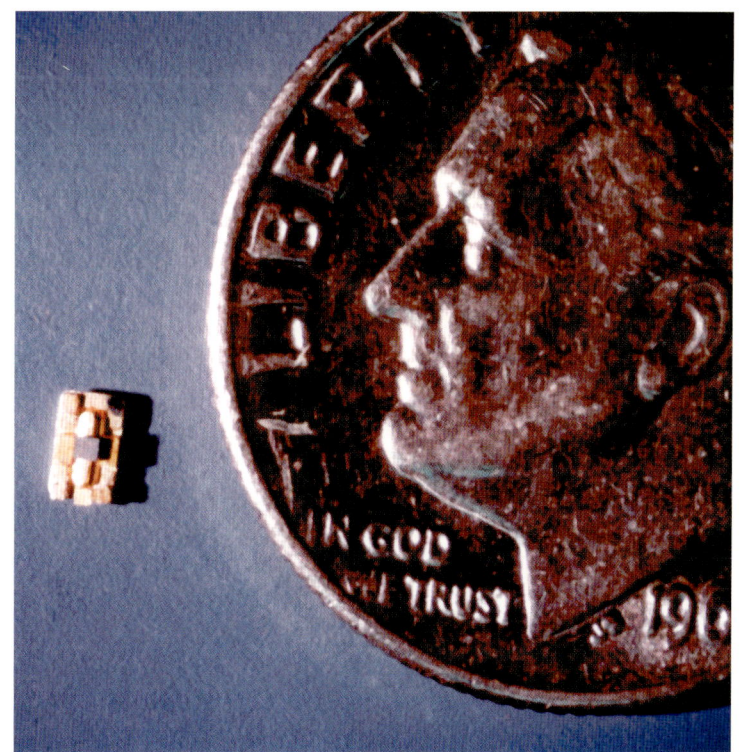

INVENTING THE FUTURE

THE MICROELECTRONICS REVOLUTION

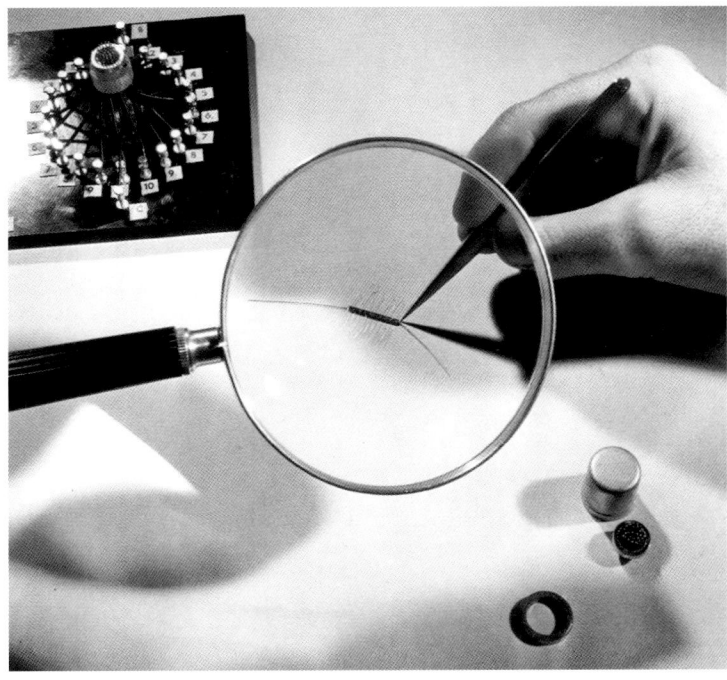

ABOVE LEFT: Equipment incorporated into a Chrysler car for testing RCA's 1802 microprocessor in 1976. The 1802 supplied electronic ignition controls in three model years of Chryslers and Cadillac Sevilles from GM.

ABOVE RIGHT: 1958 photo of the Labs' integrated circuit (IC) shift register, incorporating active and passive electronic components shown upper left. Torkel Wallmark and his staff lost the race to create the first working microprocessor to Robert Noyce and John Kilby. In the 1960s RCA continued to pioneer techniques and devices fundamental to the microelectronics revolution.

LEFT: Steven Hofstein holds up the first MOSFET circuit in 1964, just two years after he explained the technology in his lab notebook.

FAR LEFT: Eugene McDermott holds the 1802 CMOS microprocessor, which controlled electronic ignition in the Research Safety Vehicle in 1976. A number of auto safety and efficiency features in use today were developed at the Sarnoff labs.

INVENTING THE FUTURE

LEFT: Sarnoff technician places a wafer in an epitaxial deposition device.

BELOW LEFT: Art Stoller probes an integrated circuit under the microscope in 1965.

BELOW RIGHT: Rakesh Kabra and Laura Housel of Sarnoff's Microelectronics staff review the design of a new chip in 1995.

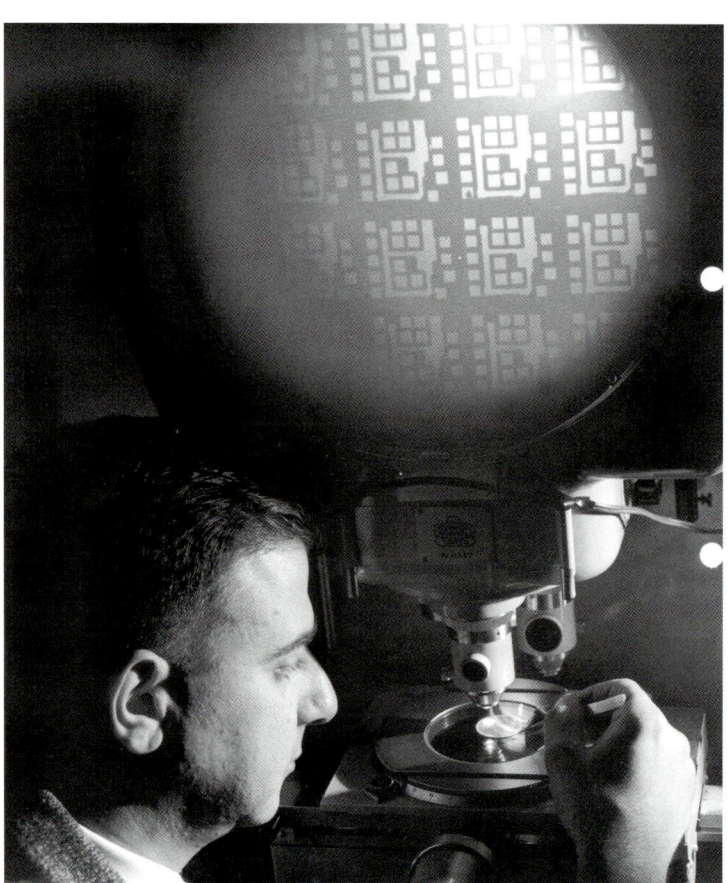

MICROELECTRONICS REVOLUTION

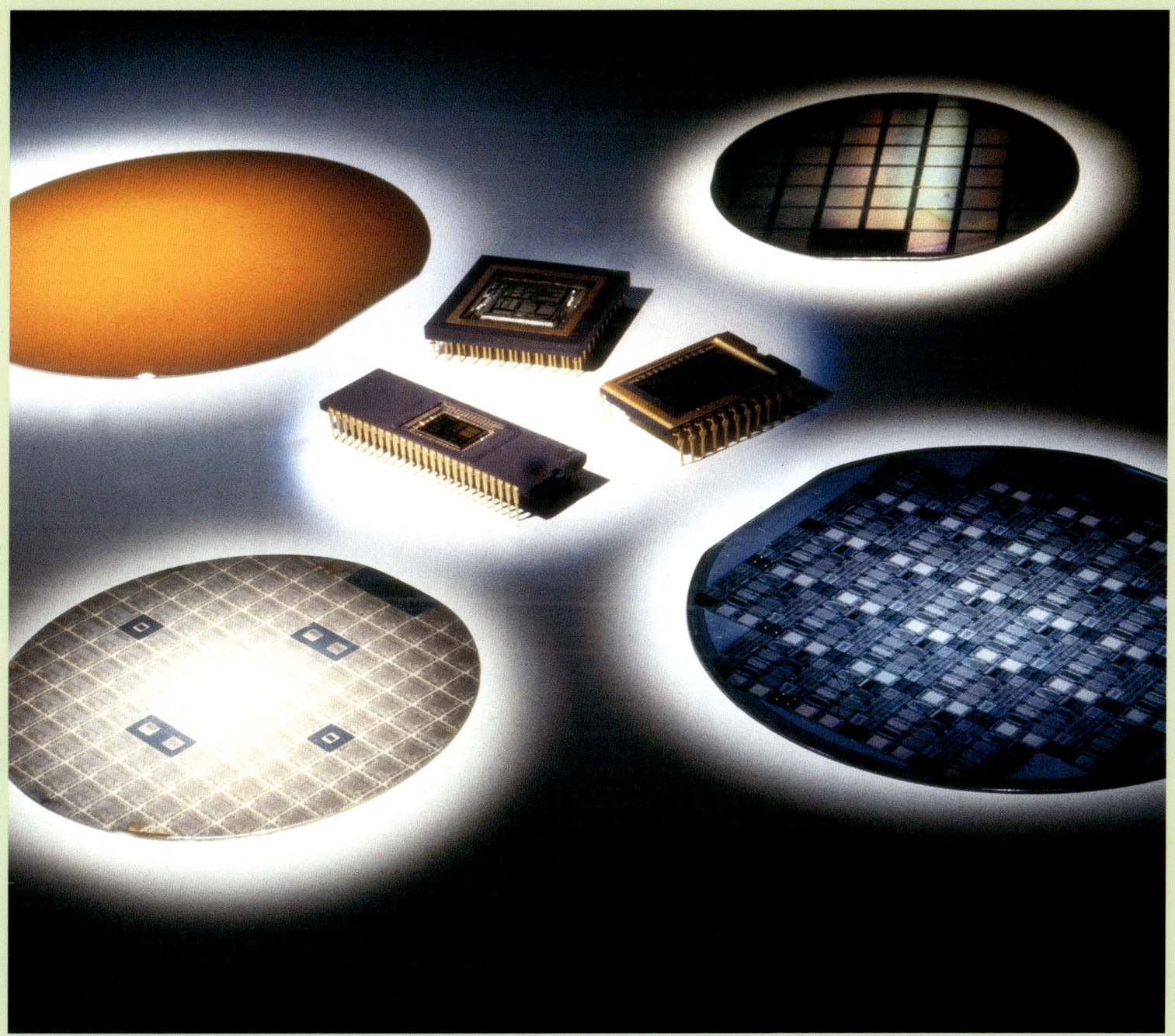

ABOVE: Wafers for imager chips and the finished imagers. Sarnoff's pioneering work includes CCD mass production, broadcast quality CCDs, back-illuminated CCDs for high sensitivity, and uncooled MEMS IR imagers. It designs and fabricates CCD, CMOS, and CMOS/CCD imagers and arrays covering the deep ultraviolet, visible, and infrared spectra.

INVENTING THE FUTURE

ABOVE: Photolithograpy technician David Stout shows a mask for a chip in Sarnoff's highly productive GEM program, c1995.

RIGHT: Robin Dawson and Steve Connor review a CCD chip design, c1995.

THE MICROELECTRONICS REVOLUTION

INVENTING THE FUTURE

Information on a Beam of Light

LASERS & LEDs

Before 1962 lasers were impractical.

Then Sarnoff researchers announced a light-emitting diode with 100% efficiency in converting electrons to photons.

Before 1977 semiconductor laser devices were too fragile for telecommunications. Then Sarnoff announced commercial laser diodes that lasted more than three years (over 25,000 hours) in continuous room temperature operation.

Today the whole world gets enormous volumes of voice, video, and data over optical networks powered by pulses of coherent light from semiconductor diode lasers.

That's just part of the story. Semiconductor light emitters and detectors preserve and disseminate cultural artifacts such as music, movies, and books on CDs and DVDs. They power scanners, printers, and data storage systems. They sculpt human corneas to correct vision, and are transforming medical diagnostics and treatment.

Sarnoff played a major role in converting lasers from lab curiosity to commercial force. Herbert Kroemer, later a Nobel laureate, worked on the basic heterojunction materials here in the 1950s. During the 1960s Sarnoff scientists solved the problems of device performance and longevity, created improved manufacturing processes, and in 1968 developed the first practical heterojunction laser (commercialized in 1969).

The pace of innovation continues. Sarnoff's unique research and production facilities keep it in the forefront of optoelectronic developments.

LEFT: Karl Hernqvist holds up the first high-powered laser to operate in the ultraviolet part of the spectrum in 1966.

INFORMATION ON A BEAM OF LIGHT

SIGNIFICANT SARNOFF OPTOELECTRONICS INNOVATIONS

1966 Development of the liquid-phase epitaxy manufacturing process for mass production of lasers, the standard process around the world today.
1967 Identification of crystal and structural defects as causes of poor laser reliability.
1968 Invention and prototype of single heterojunction AlGaAs/GaAs (pulsed) laser structure, world's first commercially viable laser diode.
1969 Commercialization by RCA of the single heterojunction laser, suitable for reliable room-temperature operation.
1971 The first visible-emission single and double heterojunction lasers emitting in the red spectrum, paving the way for optical disc, printing, and scanning systems.
1971 Development of the Large Optical Cavity (LOC) laser, the basis of today's structures.
1972 World's first blue light-emitting diode (LED).
1975 First detector-quality InGaAs material.
1975 Announcement of the first commercial-class, continuous wave, room-temperature lasers for fiber optic communications, with 10,000 hours of continuous operation.
1977 Achievement of routine laser lifetimes of over three years (25,000 hours) of continuous operation, making reliable fiber optic communications feasible.
1977 First dielectric mirrors for semiconductor lasers.
1980 Invention of high power channeled-substrate diode laser.
1982 First optical disc recording and playback with semiconductor lasers.
1984 Development of grating surface emitting high power laser.
1987 Development of 2D array for pumping solid state lasers.
1988 First GaAs-on-silicon materials growth.
1990 First 810 nm single wavelength laser.
1992 Development of integrated high-power semiconductor amplifiers.
1994 First visible distributed feedback diode (DFB) laser.
1996 First 2.7-micron-wavelength diode laser.
1997 World's highest average output power 1.5μm diode laser.
2001 Development of adjustable photonics crystal nanofluidic lenses for optoelectronics.
2003 Invention of micro-ring-resonators for chip-scale WDM.
2004 World's first 2.8μm device.

GLOBAL IMPACT

The earliest applications for Sarnoff's lasers were in ranging systems for military use. When continuous-operation lasers became available, they sparked multi-mode fiber communications (part of the $1.5 trillion global information industry), optical disc recording, and a huge market in office printers, among many other applications.

Lasers are inexpensive and ubiquitous. Communications and data storage are their largest markets, but they're also found in countless lecture pointers. They remove birthmarks, and bring 20/20 vision to the nearsighted. Lasers paint holographic images across stadiums and parks as nighttime entertainment. Lasers and detectors read data and play music and video from CD and DVD discs, and write to recordable media. High-power lasers cut fabric for clothes, or steel for cars.

Total sales of lasers, LEDs, and photodetectors in the U.S. alone are estimated at nearly $10 billion per year. Another spur to growth is the advent of blue and green lasers. These devices open up new applications in automotive and video displays, signage, even traffic lights, where they are replacing regular bulbs. The blue laser also boosts the storage capacity of an optical disc by a factor of eight.

The invention of lasers created new industries and revitalized old ones, including telecommunications. Optoelectronics-related manufacturing employs hundreds of thousands of people around the world. Its products improve the quality of human life by expanding access to knowledge, entertainment, and health services.

RELATED TOPICS

Electronic Materials: see *Elements of Innovation*
Semiconductors: see *The Microelectronics Miracle*

ABOVE: Henry Kressel, now Sarnoff's chairman of the board, with the semiconductor laser he designed for use in fiber optic communications in 1980.

INVENTING THE FUTURE

INFORMATION ON A BEAM OF LIGHT

LEFT: As part of a research project to develop atomic-level holographic data storage in crystals, a laser beamed through a lithium-niobate crystal in 1971 reveals an image of the David Sarnoff Research Center.

RIGHT: RCA's 1980 optically rewritable laser disc can store over 6GB on a 12-inch side.

BELOW LEFT: Working with Ed Miller, Jacques Pankove demonstrated the world's first blue light-emitting diode using gallium nitride in 1972, drawing on GaN photoluminescence research started by Herbert Maruska to fulfill James Tietjen's idea of a color, flat-panel LED display.

BELOW RIGHT: Patricia Cullen points to the crystal cube a laser is reading to show a map of Mercerville, NJ in 1971.

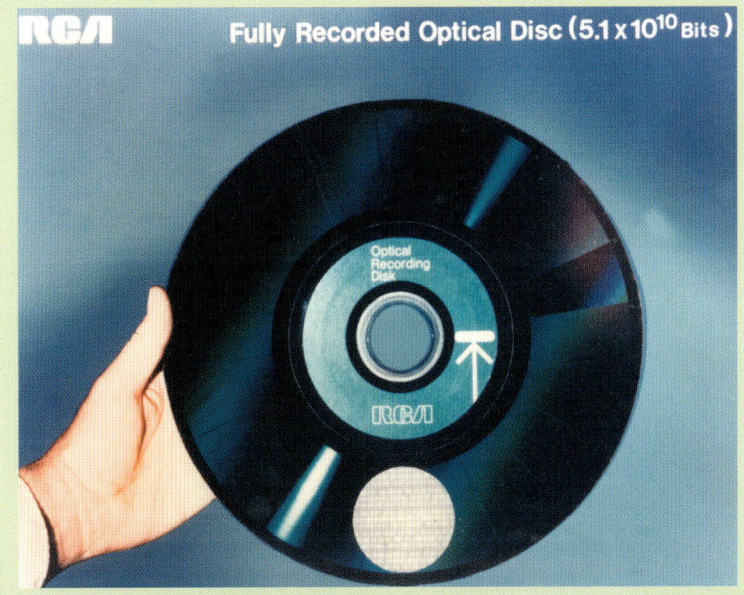

INVENTING THE FUTURE

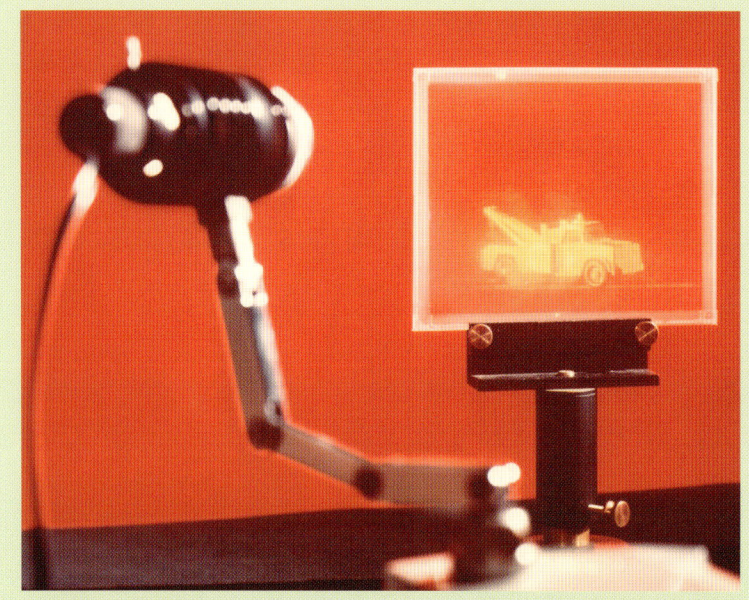

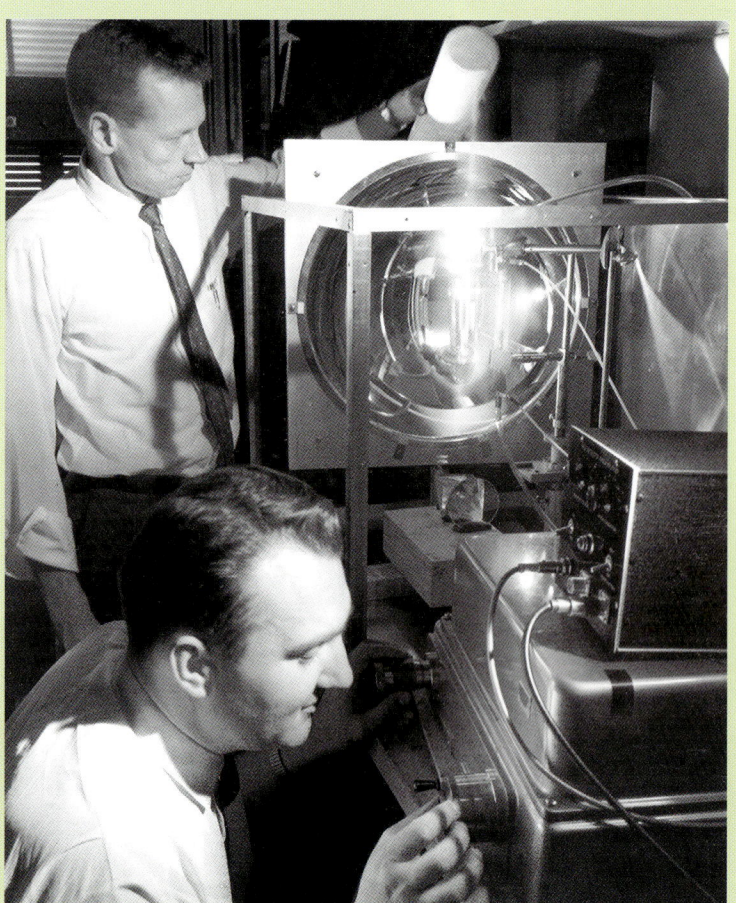

INFORMATION ON A BEAM OF LIGHT

ABOVE: James Tietjen holds a gallium-arsenide (GaAs) wafer made with the vapor-phase epitaxial growth technique that he pioneered at the Labs in 1966.

LEFT (CLOCKWISE): 1) Laser hologram, 1972. **2)** Robert Bartolini oversees laser research that ultimately leads to breakthroughs in optical recording. **3)** Robert Duncan and Zoltan Kiss measure output of a calcium-fluoride laser doped with dysprosium in 1962. **4)** Don Carlin demonstrates optical recording on a 14-inch disc in 1984.

INVENTING THE FUTURE

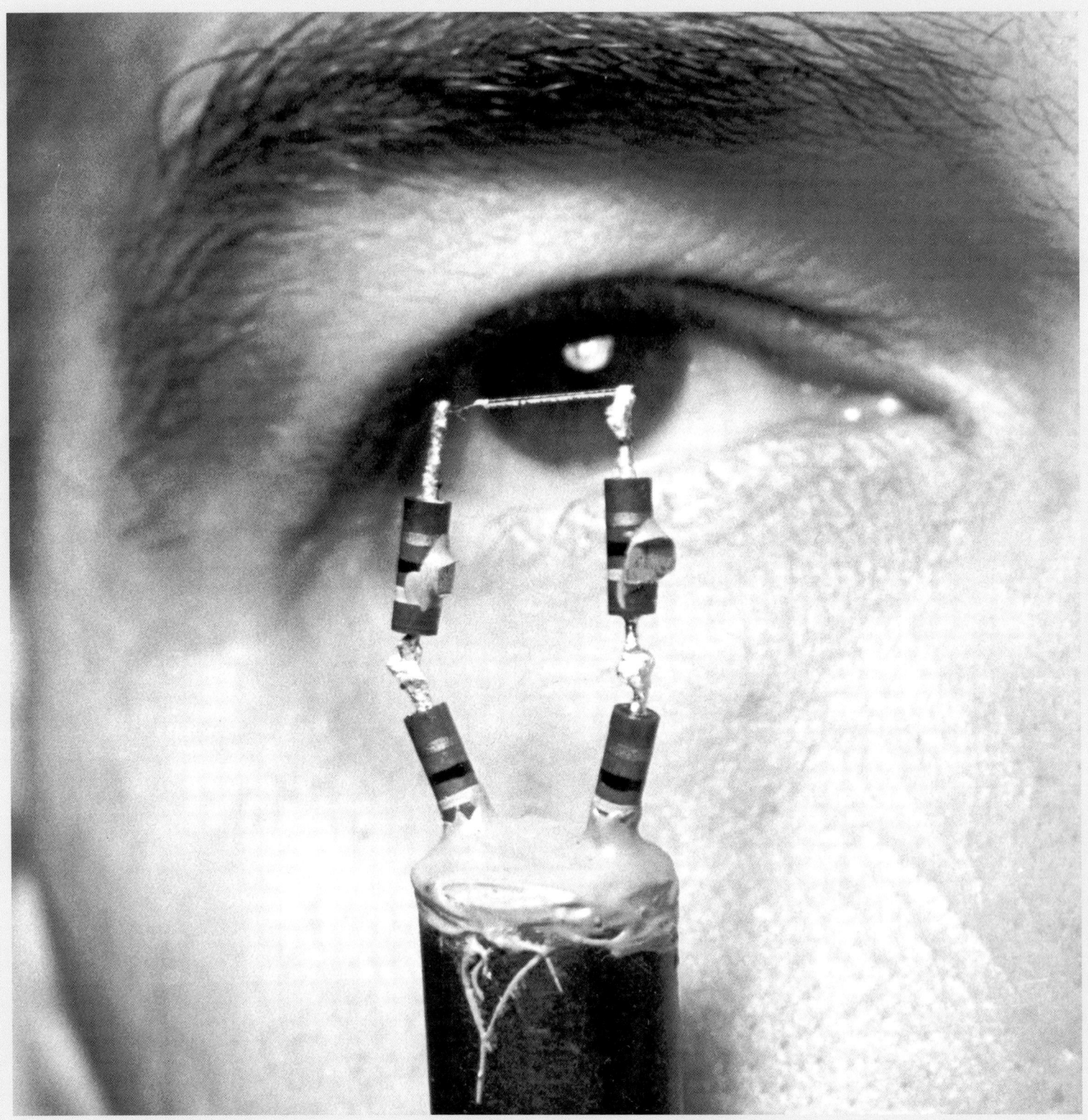

INFORMATION ON A BEAM OF LIGHT

LEFT: GaAs light-emitting diode, c1966.

RIGHT: Michael Ettenberg shows a solid-state laser with the multilayer reflector that he invented in 1978. By nearly doubling the laser's output, the reflector extended the laser's operating lifetime while reducing its power consumption.

BELOW LEFT: Nobel laureate Herbert Kroemer began his career at Sarnoff in 1954. In his three years in Princeton Kroemer commercialized the drift transistor and published his first article on the benefits of wide-bandgap heterogeneous semiconductors.

BELOW RIGHT: Grating surface emitter array, c1989.

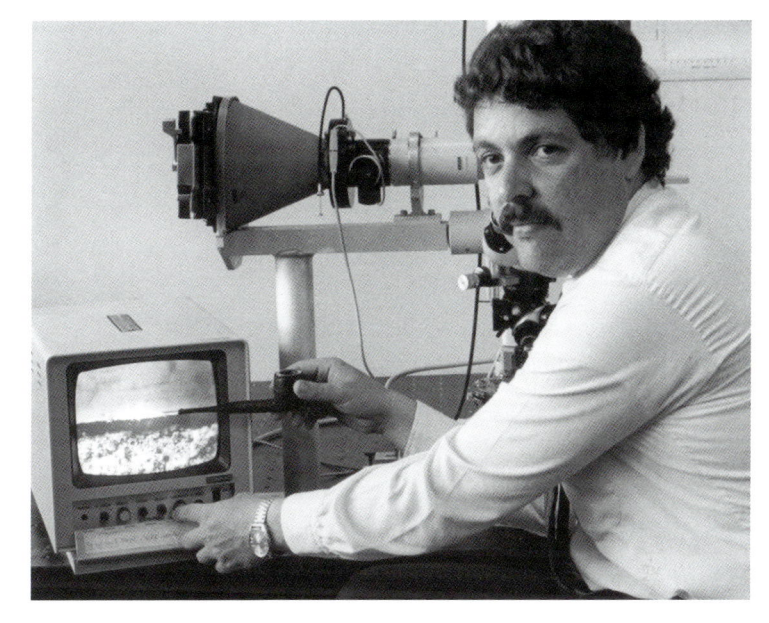

INVENTING THE FUTURE

INFORMATION ON A BEAM OF LIGHT

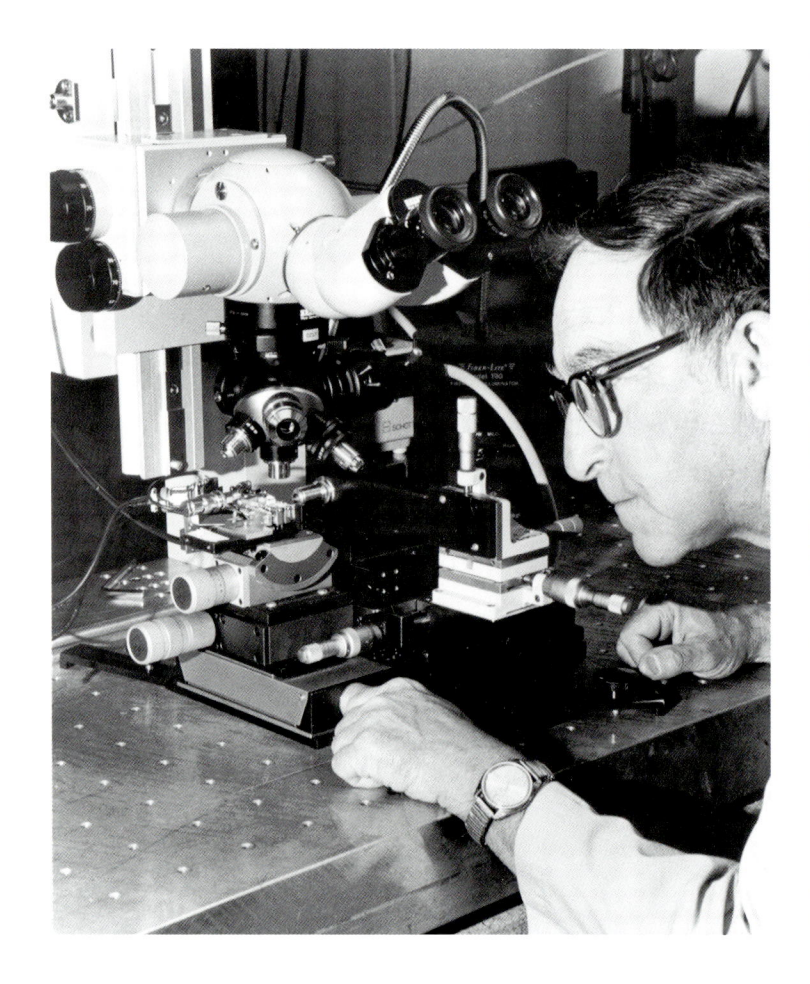

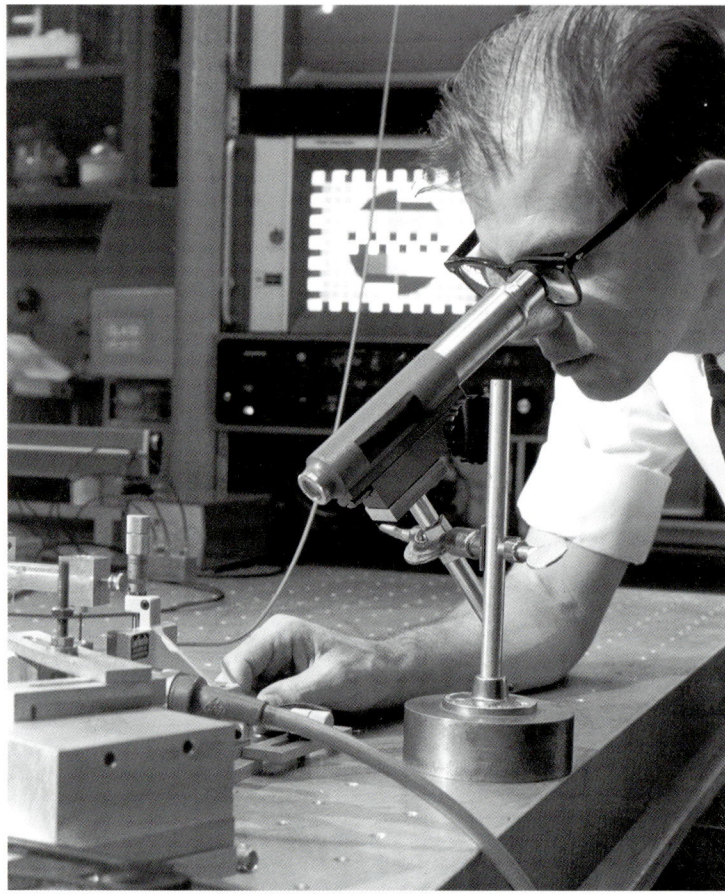

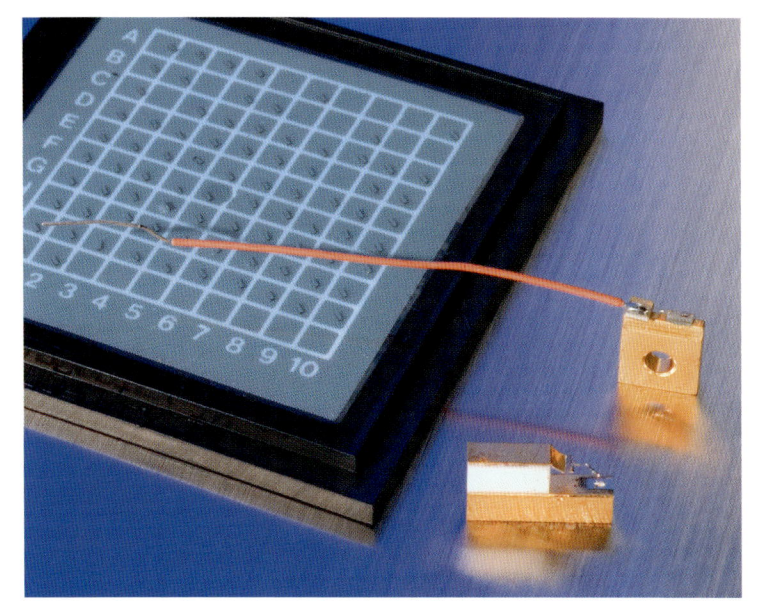

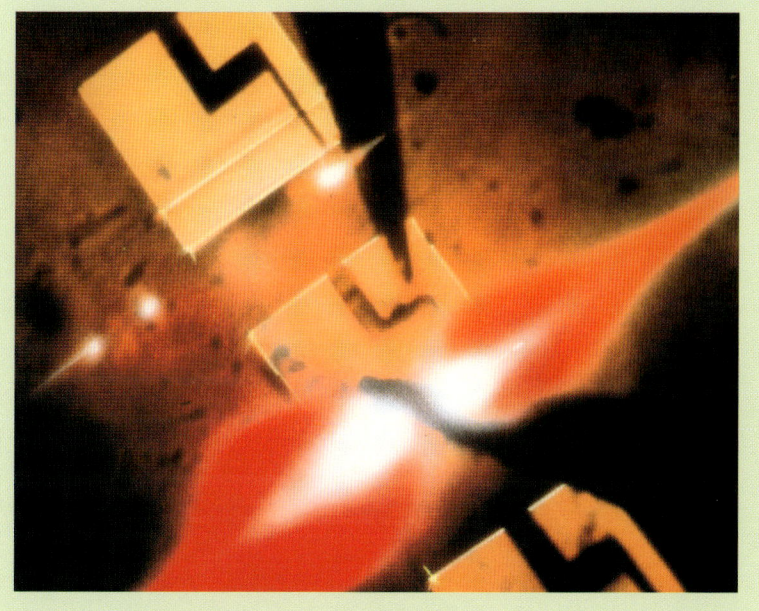

LEFT PAGE (CLOCKWISE):
1) Jacob Hammer examines an external Bragg reflecting laser in 1985. The device simplified the production of single-wavelength lasers used in space and fiber-optic communications.
2) James P. Wittke adjusts the coupling of an LED to a fiber-optic cable during experiments on the quality of analog television signal transmission over fiber optics in 1973.
3) Close-up of semiconductor laser diode chips on the tray in photo at left.
4) Tray of laser diode chips, ready for mounting in headers (right) to create laser assemblies, 2004.

ABOVE: Aluminum Gallium Arsenide (AlGaAs) laser diode is smaller than the eye of a needle.

LEFT: A laser diode emits a single-frequency red beam during testing.

INVENTING THE FUTURE

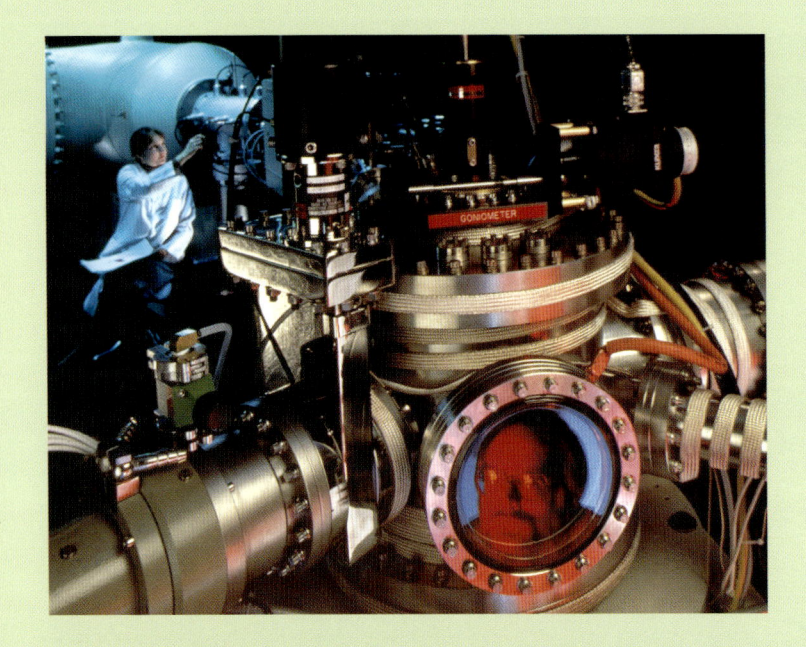

LEFT: Lori Hewitt adjusts the Rutherford backscatter spectrometer used in thin-film analysis while Charles Magee apparently poses in the window.

BELOW LEFT: Frank Kolondra uses a 300,000 electron-volt ion implanter to improve substrate doping techniques in semiconductor wafers in the early 1980s.

BELOW RIGHT: A MEMS-based atomic clock for communications use, only 3.1 cm3, relies on Sarnoff's high-speed VCSEL lasers and alkali-vapor cells for high performance and stability through novel use of atomic-state transitions. The device was co-developed with Princeton University.

RIGHT: Sarnoff's 1989 Grating Surface Emitting Array for optical networks, processing, and printing offered 100 times more brightness at lower cost and with greater reliability than an individually-assembled set of diode lasers.

INFORMATION ON A BEAM OF LIGHT

INVENTING THE FUTURE

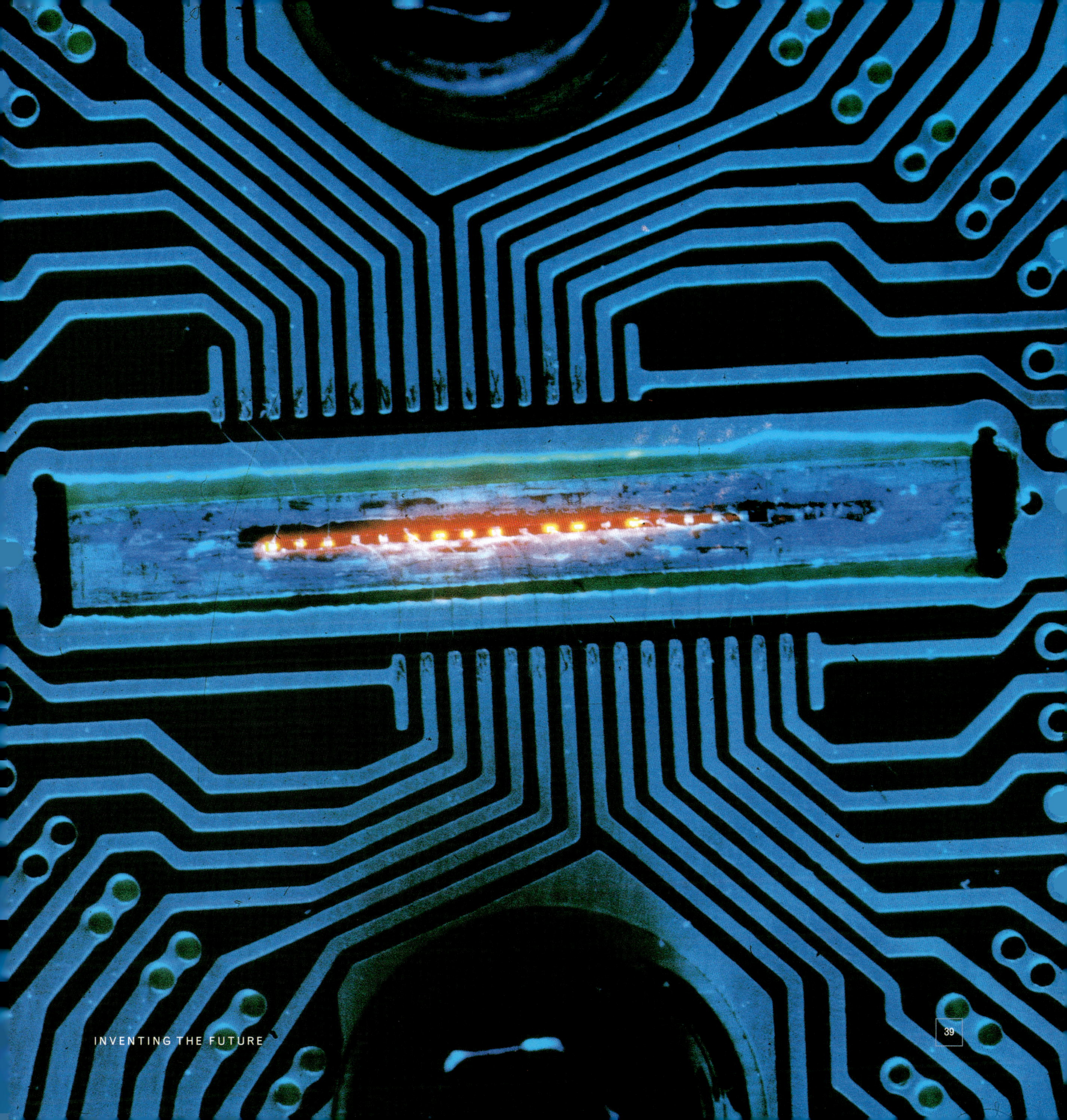

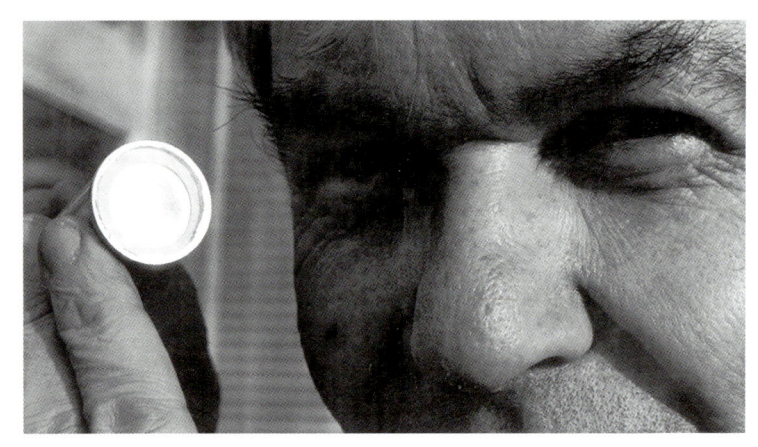

Teaching Computers to See

COMPUTER VISION TECHNOLOGY

Smart cameras make computers – and people – even smarter.

Three minutes after it's abandoned in an airport terminal, a briefcase lights up in red on a video security monitor. Is it lost luggage? Or a bomb?

Computer vision has detected it in time to find out. And not only can the security officer zoom in for a closer look — the Sarnoff-developed system will move back in time to see who left the object, and circle around that person for a better ID.

In humans the eye is an extension of the brain. Our eyes sense patterns of light, and our brains take just milliseconds to interpret the patterns as objects in a dynamic 3D world. Computer vision emulates the magic of the human visual system, using cameras and processors to play the roles of eye and brain.

This promises to transform every activity that involves seeing, analyzing, and reacting to visible phenomena. Computer vision could automatically interpret human activities for airport security, warn drivers of imminent collisions, or control robotic vehicles.

Sarnoff scientists pioneered the understanding of human vision and the development of computer vision systems. Starting with used models of neural coding, they developed a new mathematical framework called pyramid processing that lets computers do image analysis with unprecedented speed and precision.

Our systems lets soldiers inside armored vehicles see what's around them by turning their heads, as if the steel turrets were made of glass. It instantly converts aerial video into photographic terrain maps. And we have miniaturized the hardware, shrinking the processor needed for real-time vision from a dedicated computer to a single Acadia chip.

SIGNIFICANT SARNOFF VISION PROCESSING INNOVATIONS

1980 Just Noticeable Difference (JND) model of human perception, which provided a quantitative basis for designing displays and modeling vision performance.

1983 Laplacian pyramid transform introduced as basis for image compression, fusion, and enhancement; first use of wavelet technology, later adopted for MPEG-4.

1984 Pyramid Vision machine: first computer system capable of performing sophisticated visual search tasks in real time.

1991 PYR-1 ASIC shrinks pyramid-based vision processing to a single IC.

1991 VFE-100 Vision System, based on the PYR-1 and algorithms for precise electronic stabilization, allowed tracking of moving object from moving camera.

1993 First system to insert virtual advertising in live video broadcast as part of the scene.

1996 PYR-2 and VFE-200 introduced, used to perform real-time video mosaicking (creating panoramic views from video frames), stereo range analysis, and visual control for unmanned ground and air vehicles.

1996 Iris recognition at a distance introduced for user identification at ATMs and other applications.

1997 Sarnoff commercial software makes mosaicking possible with consumer-grade digital still and video cameras.

1998 Pyramid Vision Technologies subsidiary formed to market Sarnoff's high-performance vision systems.

2000 Acadia I, a full real-time vision system on a chip.

2001 Video georegistration deployed; finds objects on the ground in aerial surveillance video.

2001 "See-through turret" visualization integrates multiple cameras into one display, providing tank operators with a seamless panoramic view of the surrounding area; used on military vehicles from United Defense LLP.

2003 Video Flashlight surveillance systems completed and deployed worldwide, providing a unique ability to fly around a scene and fly back in time by combining multiple video feeds; now licensed to and marketed by L3 Communications.

2004 Silicon IP to deliver panoramic photos from standard cell phone cameras.

2005 Video Flashlight licensed for commercialization to L3 Communications.

2005 Iris on the Move™ system demonstrated — it captures images of the human iris for identification purposes as people walk through a gateway.

GLOBAL IMPACT

The plummeting cost of cameras and computing power is driving computer vision into more consumer, government, and business applications.

- **Medical**: automatic comparison of yearly MRI breast scans to spot the emergence of suspicious lesions.
- **Commercial**: turning shaky video into rock-solid imagery in real time; overlaying ads on walls or color demarcation stripes on athletic fields in TV broadcasts; inserting objects into live video.
- **Military and security**: detecting intruders in a restricted area; merging external data with panoramic live views to pinpoint target locations; unifying all security systems on a single screen
- **Consumer electronics**: seamlessly combining successive video frames into sweeping panoramic views; increasing the clarity of an image by combining the information from different frames.

Computer vision has a role in any activity where seeing and responding to visual information is important. Its full implications have yet to be realized.

RELATED TOPICS

Television and Video: see *The Video Revolution*; *The World on Display*
Cameras: see *Extending the Power of Sight*
Computers: see *Transmitting Intelligence*

ABOVE: Acadia® I vision processing chip on plug-in PC board.

LEFT: Sarnoff's silicon intensifier tube provided color video from the Apollo 15 lunar mission. It met NASA's need for an imager that would not burn out in direct sunlight.

INVENTING THE FUTURE

TEACHING COMPUTERS TO SEE

LEFT: RCA Laboratories collaborated with RCA Victor during World War II to develop the first TV-guided missiles.

RIGHT: Infrared signaling systems, based on the Labs' IR camera tubes. RCA Victor produced thousands for nighttime military operations during World War II.

BELOW LEFT: 1946 video image of the Anacostia power station in Washington, D.C. from a remotely guided aircraft or unmanned aerial vehicle (UAV), captured by the image orthicon tube invented in 1943.

BELOW RIGHT: "See-Through Turret" with head-mounted display merges views from multiple cameras to give armored vehicle operators a 360-degree view around the vehicle by moving their heads.

INVENTING THE FUTURE

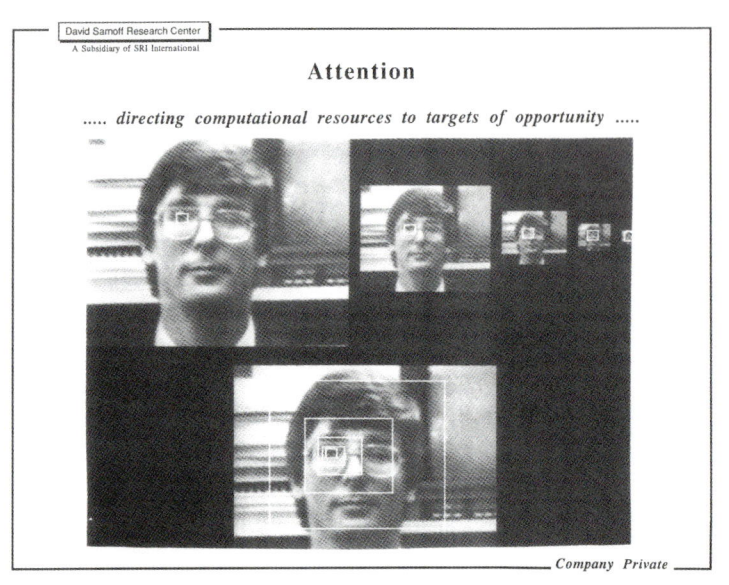

ABOVE: Ralph Klopfenstein, Curt Carlson, and Albert Pica were involved in early computer vision work in the 1990s.

BELOW: By the end of World War II RCA's video camera and display technologies were being incorporated into guided missiles.

ABOVE: Curt Carlson's photo is broken into image pyramids for computerized vision processing, then analyzed as part of a face recognition experiment.

BELOW: Sarnoff's computer vision technology can drape three-dimensional models with real-time video imagery (vehicles, people among buildings).

TEACHING COMPUTERS TO SEE

TOP: Acadia vision chip and its associated board enables real-time vision processing in a standard PC, including stabilization, motion detection and tracking, visible/IR fusion, and other functions.

INSET: *Middle left:* Sarnoff's Pyramid Vision processors can mosaic related images, as in this panoramic picture of the Philadelphia Museum of Art created from multiple frames c1997. *Middle right:* Pyramid Vision processors enable the fusion of near and deep focus imagery into a single high-resolution image c1997. *Left:* Vision processing clears up blurred video sequence, enables high-definition resolution and sharp close-up c1997.

INVENTING THE FUTURE

45

Keith Hanna tests the ability of an early computer vision robot to track motion.

TEACHING COMPUTERS TO SEE

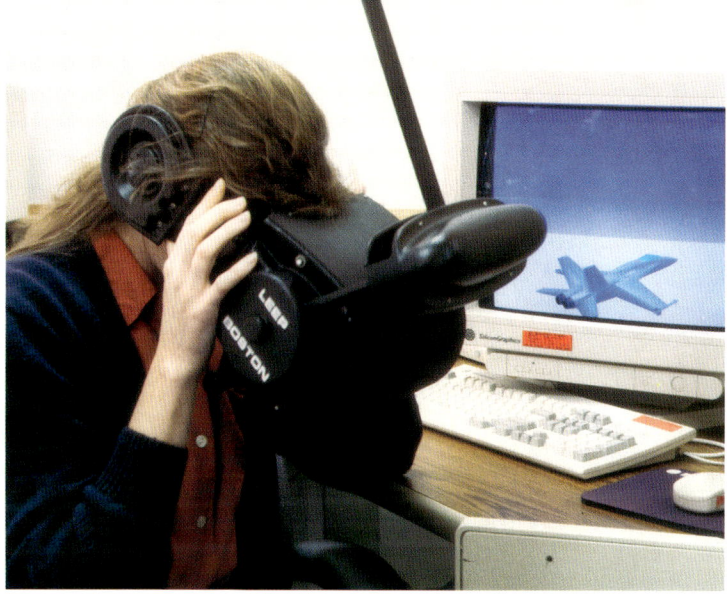

ABOVE LEFT: Jim Bergen, Charles Anderson, Edward Adelson, and Peter Burt, along with Joan Ogden (not shown), laid the base in 1984 for over 20 years of innovations in computer vision through the application of Laplacian algorithms to image pyramids.

ABOVE RIGHT: A staff member tests a prototype immersive combat flight training system, built around computer graphics in the early 1990s.

LEFT AND BELOW: The Pyramid Vision VFE-200 (background) helped UN forces during the Kosovo conflict in 1998 by processing video in real time from a moving platform. Three years later the single-board, ruggedized Acadia® I (foreground) added video fusion and other capabilities.

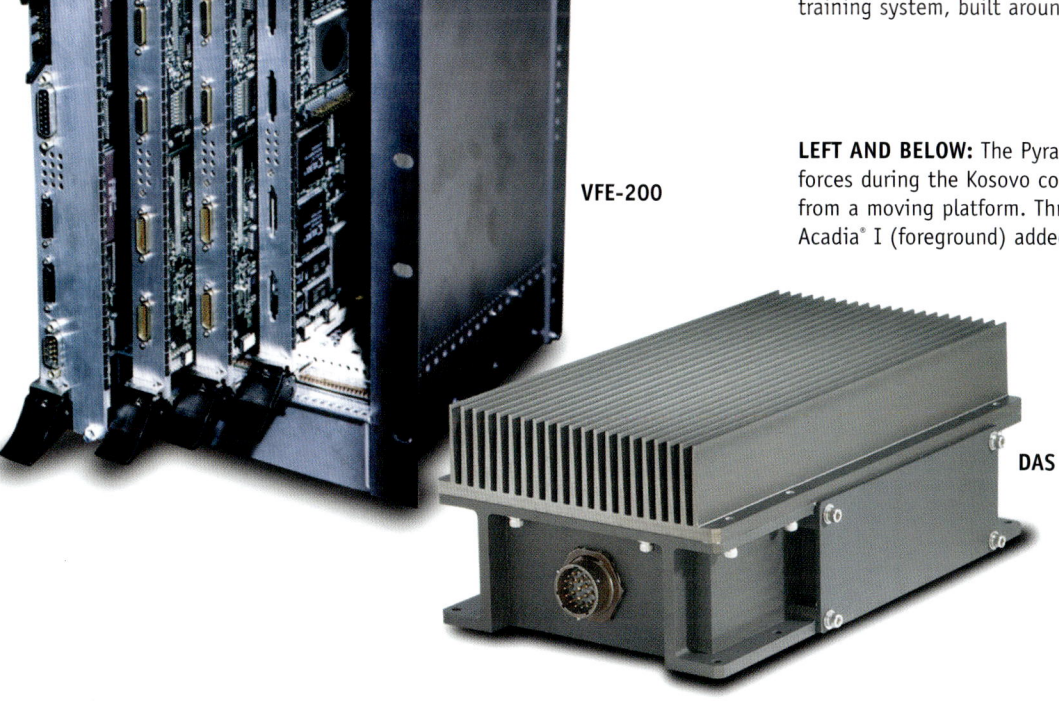

VFE-200

DAS

INVENTING THE FUTURE

The World on Display

PICTURE TUBES, FLAT PANELS, AND PROJECTORS

Video lets pictures live.

Photography freezes moments in time. Video shows them in flight, involving us in movement and action, showing us history as it happens. Displays let us see it all.

The original electronic video display was the monochrome cathode ray tube (CRT). By pressing it into service as the "picture tube" for its electronic television system, RCA turned this scientific curiosity into a consumer necessity. It changed the way society is entertained and informed.

Video displays quickly grew brighter and larger, with more realistic images. A major advance was Sarnoff's creation of the shadow-mask CRT, which made color TV practical. Flat-panel technologies such as liquid crystal displays (LCDs), also developed at Sarnoff, made displays thinner, lighter, and more portable. Projection displays have brought video into the business world, and big-screen theaters into the home.

Innovation also made high-quality displays manufacturable and affordable. In 1953 RCA's first color TV, with a 12½-inch CRT, cost as much as a full-size car. In 2005 you can buy a 50-inch rear-projection HDTV boasting three CRTs, or a powerful laptop PC with ultra-sharp video on an LCD screen, for under $1,000. Some color TVs now cost less than $100.

In addition to Sarnoff's invention of the color CRT and the LCD, our work on solid-state light-emitting diodes (LEDs) helped make large outdoor video signs feasible. Sarnoff innovators demonstrated the first deformable mirror projector, forerunner to today's digital cinema projectors. Their exploration of tiled and flexible organic displays could produce the most exciting display innovations yet.

ABOVE: Joseph Castellano peers through a dynamic-scattering liquid crystal display (LCD) shutter in 1968.

SIGNIFICANT SARNOFF DISPLAY INNOVATIONS

1950 Color shadow mask CRT, the color television picture tube and the basis for all later color tube displays.
1961 Thin-film transistors (TFTs), widely used as drivers for liquid crystal displays (LCDs).
1963 Discovery of the electro-optical properties of liquid crystals.
1965 Demonstration of the first colored LCD.
1966 Development of "dynamic scattering" LCD, the first LCD to be commercialized.
1970 Demonstration of a "deformable mirror projector," forerunner of modern micromirror or digital mirror device (DMD) projectors.
1989 Advanced polysilicon active-matrix, liquid-crystal displays (AMLCDs).
1993 Blue EL electroluminescent (EL) phosphor, the last color element needed to make a full-color EL display.
1994 5-megapixel monochrome CRT display with 200 fL brightness for high performance medical and surveillance applications.
1994 First integration of drive circuitry on an LCD display using amorphous silicon transistors.
1995 Active matrix EL display on a silicon chip, demonstrating 800 fL peak brightness and over 2000 pixels per inch resolution.
1996 High-resolution LCD projector (1280 x 1024 pixels).
2001 Demonstration of new CRT design with half the depth and twice the resolution of conventional picture tubes.
2001 Flexible plastic video-capable LCD display using organic transistors, produced on an ink-jet printer.
2002 Partnership with DuPont and Bell Labs to develop flexible high-resolution organic LED (OLED) display using organic transistors, printable on web-based press.
2004 Partnership with DuPont, others to develop rollable, flexible OLED display at lower resolutions for field use by military, emergency, other services.

GLOBAL IMPACT

Displays have had a pervasive impact on how the people around the world are entertained and access information. They are an important growth industry, with an annual market of $66.5 billion for the components alone.
• Over *three billion* shadow mask color CRTs have been sold worldwide since 1960. 95% of all TVs made today still use CRTs.
• 2.8 billion LCD screens valued at $47 billion are shipped every year.
• Over two million DMD-based projectors for business presentations and rear-projection TVs will be shipped in 2005.

Display technology lets workers interact with text, graphics, and video in real time, instead of waiting for a printout. This made the Information Age possible. The displays in medical imaging systems make diagnosis easier and more accurate. Displays that show surgical sites have enabled minimally invasive surgery. 3-D animation also relies on high-resolution display. And displays allow security personnel to keep watch on remote locations.

Whole industries continue to develop around display technology, from graphics cards to consumer camcorders and portable DVD players.

RELATED TOPICS

Television and Video: see *The Video Revolution*
Phosphors: see *Elements of Innovation*
LCDs: see *The Microelectronics Miracle*

ABOVE: An electron gun for a color television picture tube. Sarnoff staff continued to innovate in the technology 50 years after inventing it.

INVENTING THE FUTURE

THE WORLD ON DISPLAY

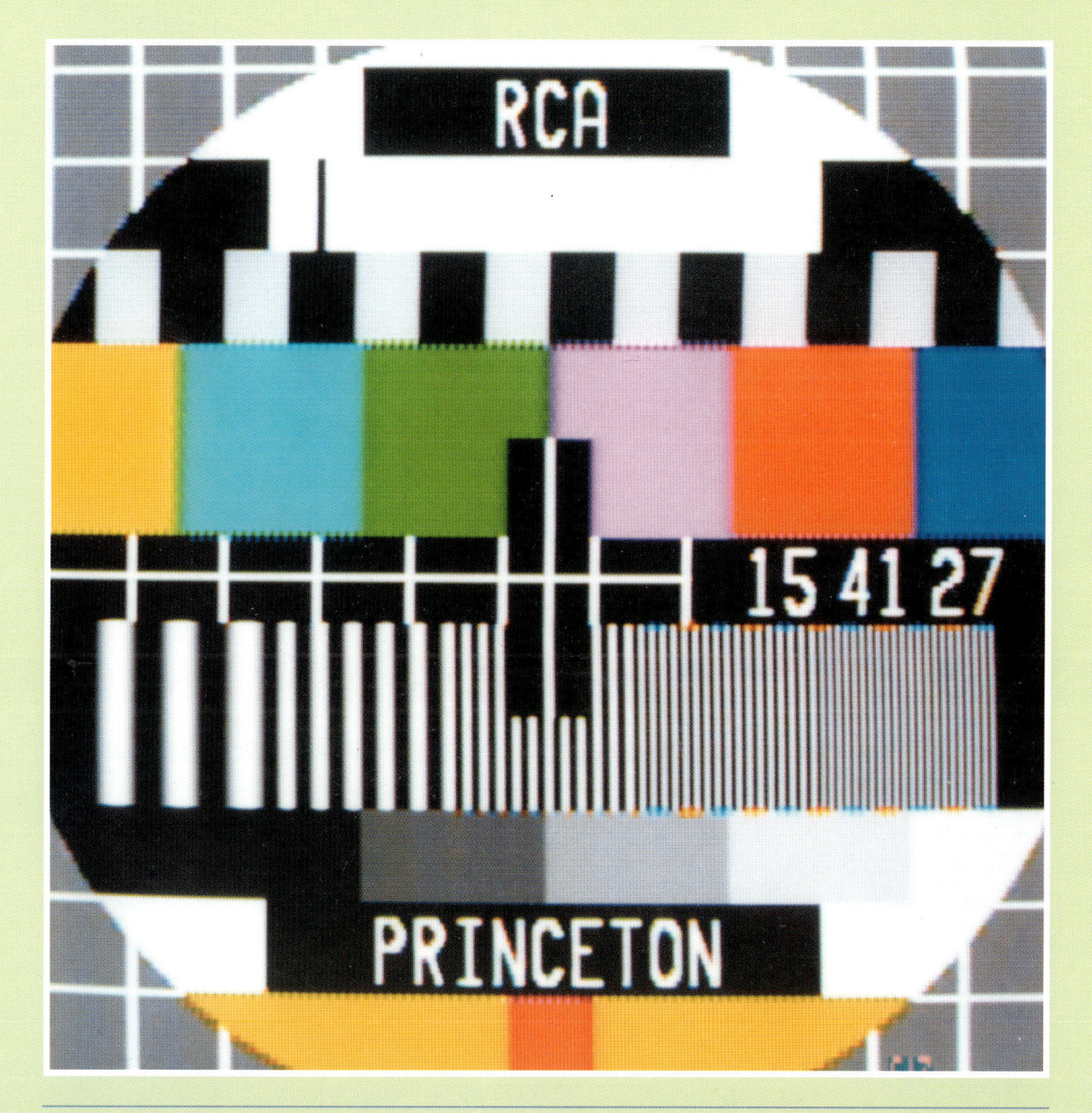

ABOVE: Test patterns such as this one, developed at the Labs, are a standard tool for measuring and calibrating the output of video systems.

LEFT: RCA's "million-proofed" 830TS receivers on the production line in Camden, NJ, in 1948. The chassis was based on the compact designs developed for guided missile control displays.

INVENTING THE FUTURE

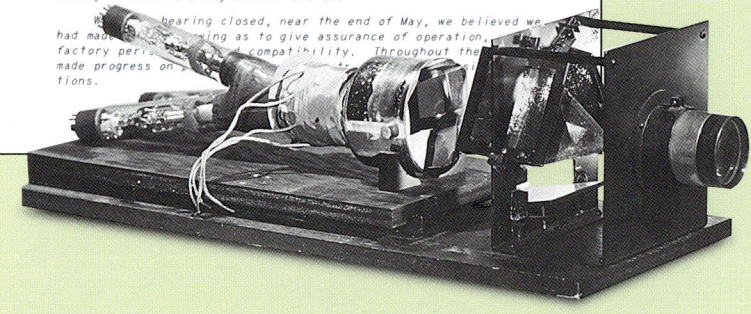

ABOVE LEFT: In 1970 RCA championed the liquid crystal display (LCD), invented at the labs, for time-keeping applications.

ABOVE: The shadow-mask cathode ray tube (CRT), the first practical color video display, emerged from a six-month effort at the Labs. Combining red, green, and blue in a single-gun device, it was the basis of all TV and computer picture tubes.

LEFT: In response to David Sarnoff's request for a flat-panel television, Labs staff demonstrated a "light amplifier" based on electro-luminescent projection in 1956.

RIGHT: A 1946 concept photo promoting electronic color television against a rival electro-mechanical system required viewers to look past the tape holding a color transparency on the screen.

THE WORLD ON DISPLAY

INVENTING THE FUTURE

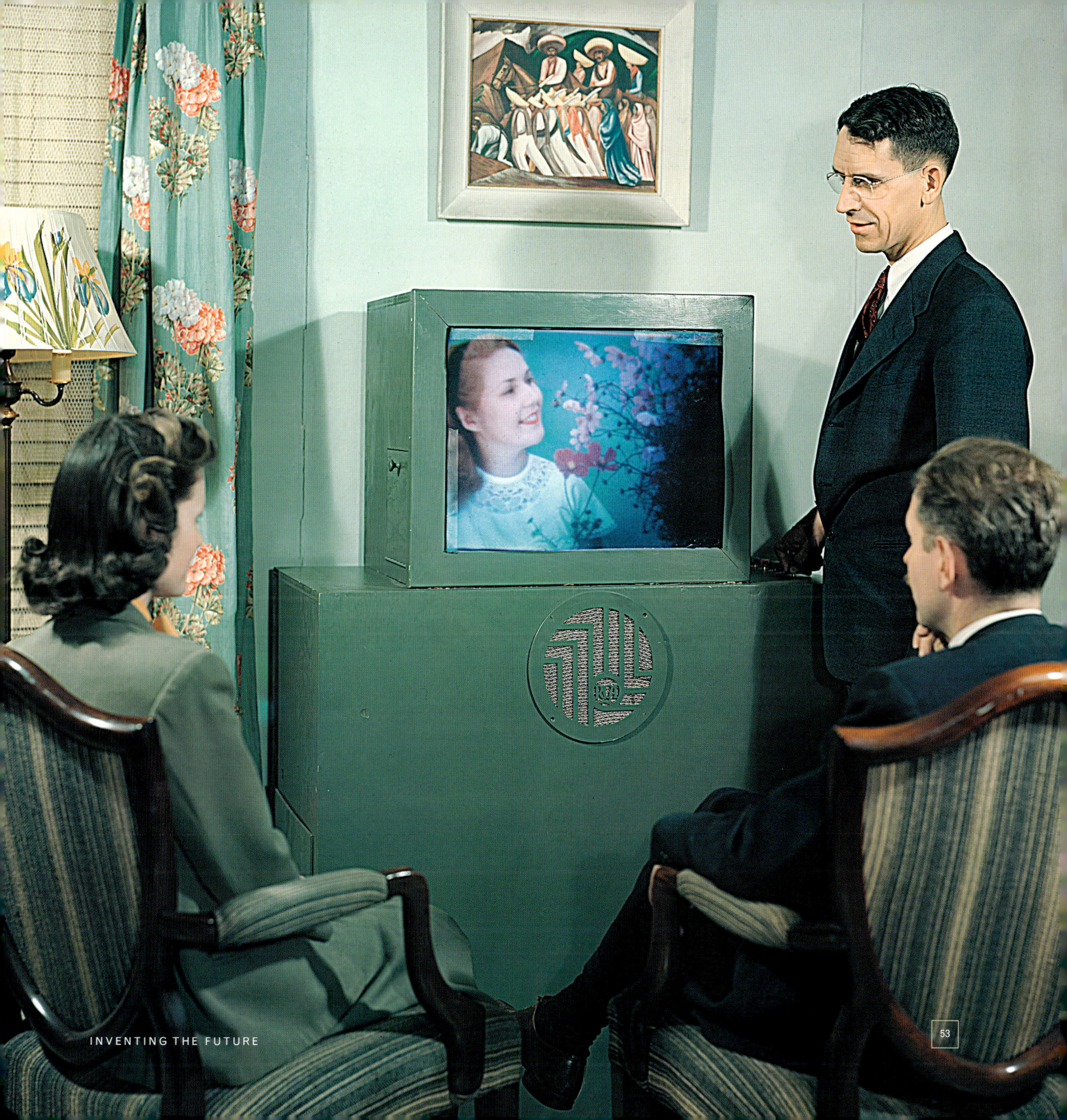

ABOVE: George Heilmeier won the Kyoto Prize in 2005 for leading the team that designed, built, and demonstrated the world's first liquid crystal displays between 1965 and 1968.

RIGHT: RCA Labs Director Elmer Engstrom points to the winner of the Labs' internal competition for a practical color video tube in 1950, as (l to r) project director Edward Herold, shadow-mask inventor Harold Law, and electronic television pioneer Vladimir Zworykin look on.

THE WORLD ON DISPLAY

ABOVE: Polysilicon active matrix LCD, built for Wright-Patterson Avionics Laboratory, U. S. Air Force in 1989, was the world's largest such display for 10 years.

ABOVE RIGHT: A Sarnoff-designed electromagnetic yoke for the CRT in a wide-screen TV, which deflects the electron beams that "paint" the video images without distorting the picture.

RIGHT: World's first video-capable flexible plastic LCD. The display, printable on an ink-jet device, was jointly developed by Sarnoff with Pennsylvania State and Kent State Universities in 2001.

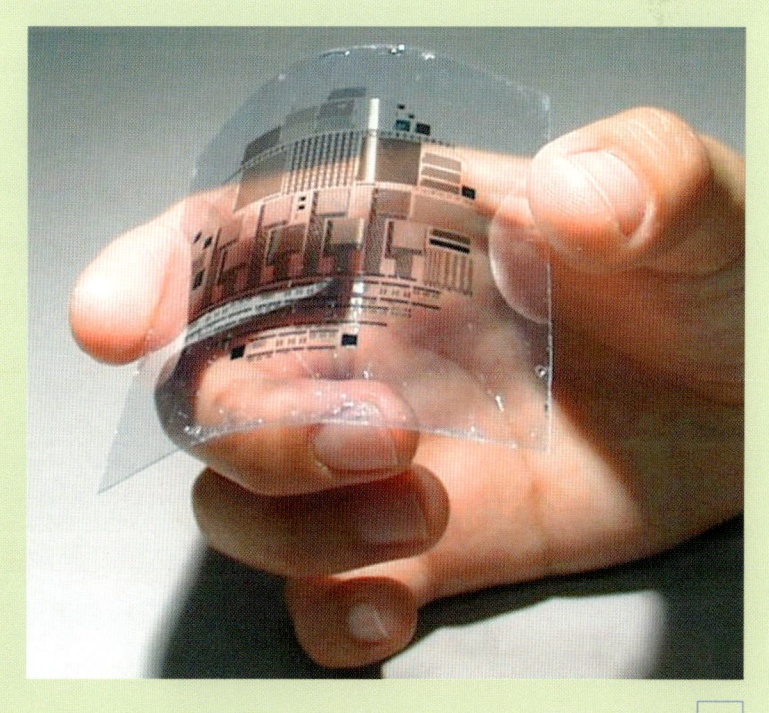

INVENTING THE FUTURE

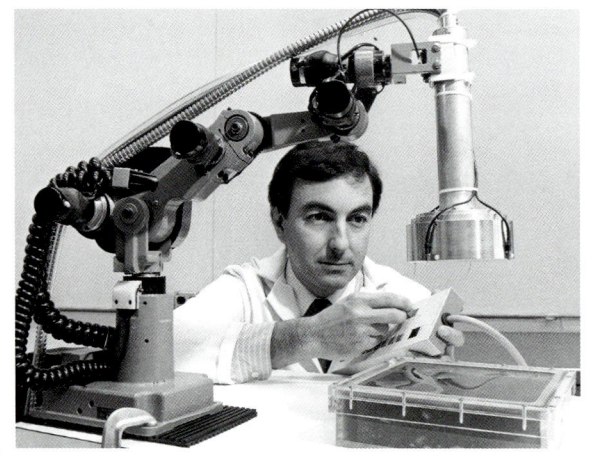

Robert Paglione controls a hyperthermic microwave applicator in 1986. Sarnoff helped pioneer microwave treatment of tumors beginning in the mid-1970s.

The Bionic Future
DIAGNOSES, DRUGS, DEVICES

People are living longer, healthier lives.

In the last 200 years the average human lifespan in the industrialized world has doubled, to 70 years.

But this is just the beginning. Advances in biology, genomics, and electronics promise faster, more fundamental progress in bettering the human condition.

Sarnoff has been an active participant in this vital effort for over 60 years, even when biomedical innovation was not a core part of its mission.

In the early years its innovations included new analytical equipment, drug manufacturing methods, and patient treatment technology. Sarnoff developed new microscopes to reveal the processes of life. It found ways to boost production of a lifesaving antibiotic, and created electronics for medical lab work.

In 1992 the time was right for the company to expand its technology portfolio and create a new biomedical group. Advances in genomics, combinatorial chemistry, manufacturing technology, computational chemistry, and microtechnology were transforming pharmaceutical and biomedical technology. Sarnoff's strengths in electronics, engineering, and physics gave it an advantage in this environment.

Since then Sarnoff has created better ways to develop and deliver effective drugs. It has pioneered vision-based cancer detection, high-throughput drug screening, and disposable healthcare devices. It has founded successful ventures in these fields, usually in partnership with pharmaceutical companies. And new, faster ways to detect, identify, and counter biothreats before they can wreak havoc are currently in development.

SIGNIFICANT SARNOFF BIOMEDICAL INNOVATIONS

1940s
- First practical transmission electron microscope.
- Radio wave method for dehydrating penicillin (1943), forerunner of microwave drying.
- Electronic OCR-based reading system for the blind.

1950s
- Video microscope.
- Video-based high-speed blood counting device (sanguinometer).
- Ingestible FM "radio pill" to transmit conditions inside the body of a patient.
- Ultraviolet color-translating microscope for viewing live microorganisms without staining (and killing) them.

1960s
- Miniaturized solid-state "radio pill."
- Threshold learning process for computerized feed-back adaptation based on Markov chains.

1970s Demonstrations of ultrasound imaging.

1980s Microwave-based hyperthermic applicators to treat tumors.

1994 MicroLab Diagnostic: micro-laboratory concept for a completely automated MEMS-based instrument to analyze micro-volume blood samples for a broad range of diseases.

1995 Ultra high speed DNA sequencing using SEQ's single model detection technology to sequence and store personal genome information on a CD-ROM for diagnosis and genomic therapy.
Development of MicroLab-like on-chip combinatorial chemistry for high-throughput drug development.

1997 New approaches to very accurate delivery of drugs through inhalers or transdermal patches, and for manufacturing precise tablet dosages via dry powder deposition.

1999 Vision-based software that does automated comparison of MRI scans for early detection of breast cancer.

2000 Sarnoff venture company Songbird markets world's first disposable hearing aid, developed by Sarnoff.

2001 Development of a revolutionary method for de novo drug design on a cluster computer.

2004 Automated insulin pump for diabetics.

GLOBAL IMPACT

The biomedical industry is the most likely source of future improvements in the length and quality of human life. Pharmaceuticals, healthcare products and services, diagnostic technology, and over-the-counter drugs also constitute a major part of the world economy.

World healthcare sales are currently estimated at over $3 trillion per year. The annual figures from the U.S. are just as impressive:

- Ethical pharmaceutical revenues are well in excess of $200 billion, with biotech-based therapeutics accounting for $20 billion and growing at over 10% a year.
- Sales of surgical devices have reached nearly $50 billion/year.
- Sales of diagnostic devices exceed $13 billion, with one-third of that accounted for by diagnostic imaging and the rest by in vitro devices.

Sarnoff's contributions are just beginning to bear fruit. In pharmaceuticals there is a demand for faster, more accurate identification of promising compounds to fight disease. The Sarnoff-developed technology at Locus can identify drug leads in silico 80% faster.

As surgery becomes less invasive, the ability to image what is happening at the surgical site becomes more important. Sarnoff vision technology promises important advances, to complement its utility in diagnostic imaging.

In healthcare, Sarnoff's skill at miniaturization has already produced a disposable hearing aid, and its wireless expertise promises an in-the-ear life signs monitor that hospitals and physicians can use to keep tabs on ambulatory patients. Coatings for surgical instruments are another promising area.

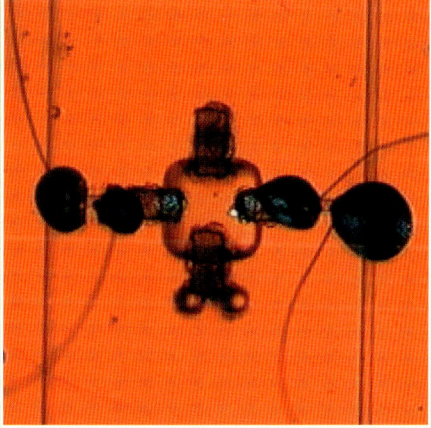

RELATED TOPICS
Semiconductors: see *The Microelectronics Miracle*
Cameras and imagers: see *Extending the Power of Sight*
Vision technology: see *Teaching Computers to See*

ABOVE: Microphotograph of one reaction cell with its tiny pumps and vias, part of a chemistry lab on a chip developed by Sarnoff in 1996.

THE BIONIC FUTURE

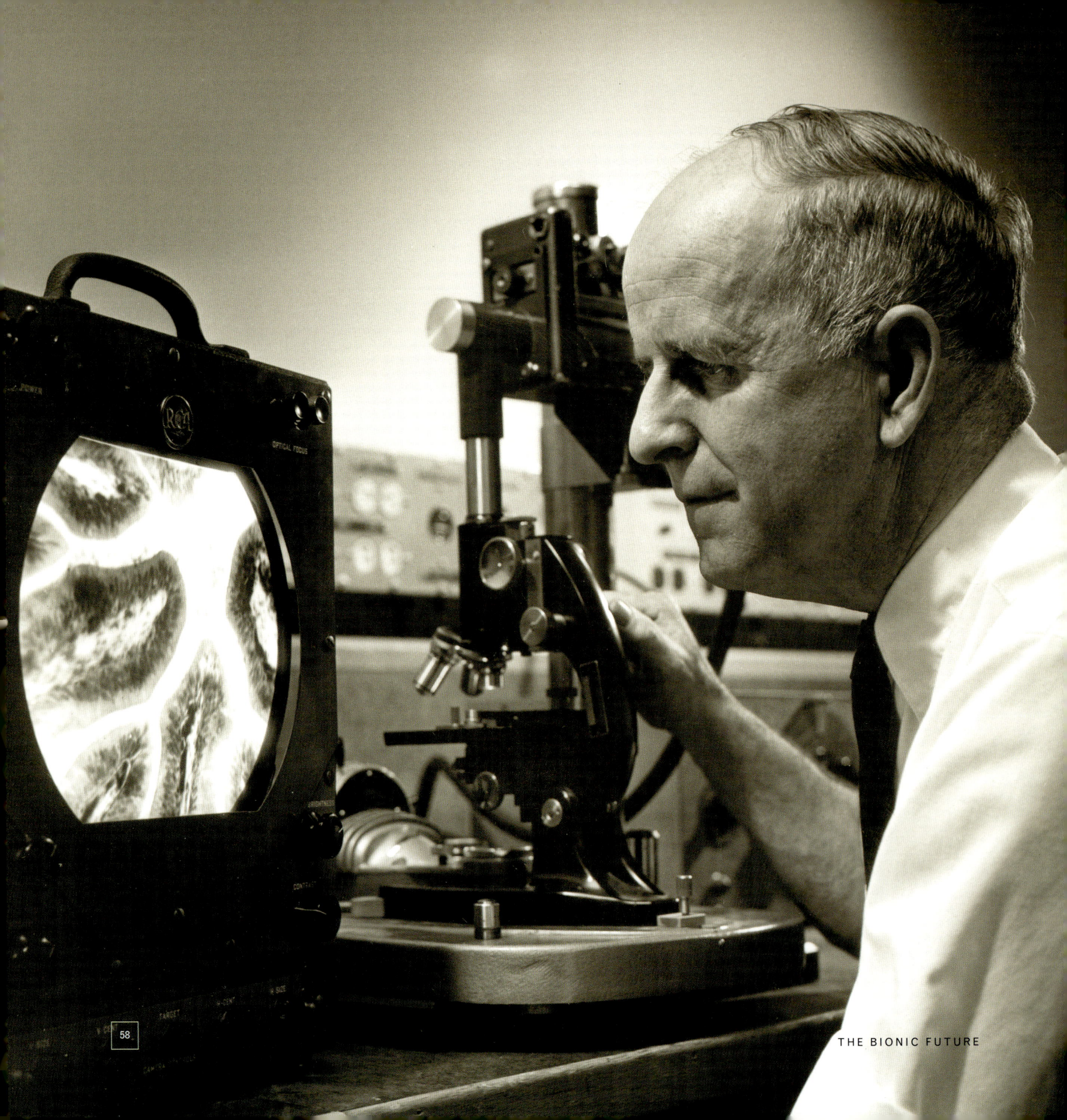

ABOVE: James Hillier shows Vladimir Zworykin a table-top version of his electron microscope during World War II. Hillier developed the first commercial electron microscope in 1940 at RCA Victor.

LEFT: Les Flory views color video of red blood cells in 1957 as part of a project to develop a sanguinometer.

INVENTING THE FUTURE

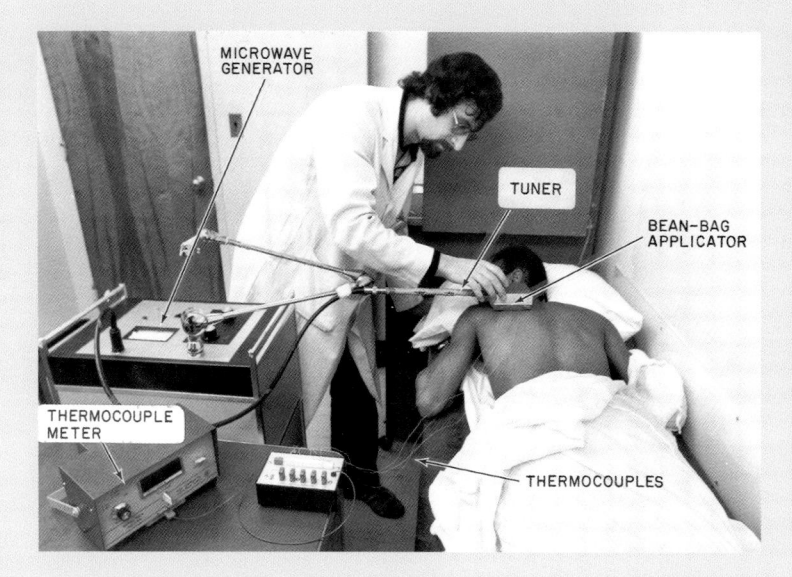

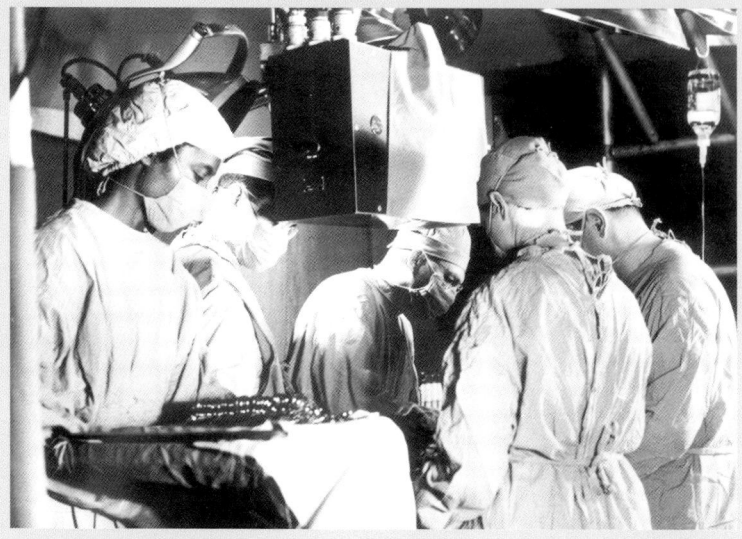

ABOVE LEFT: 1976 application of a hyperthermic microwave system for non-invasive treatment of a tumor, pioneered by Sarnoff.

ABOVE RIGHT: A surgical team in Philadelphia operates under an RCA color vidicon camera that enables remote viewing of the procedure for training purposes.

LEFT: Bacillus cereus magnified 62,500 times by an RCA electron microscope. After developing the device, James Hillier spent ten years at the Princeton Labs refining the technology and collaborating with doctors and biologists on best practices.

RIGHT: George Brown shows off the first system for bulk dehydration of penicillin by radio frequency heating during World War II. E.R. Squibb bought and operated a descendant of this technology, enabling the mass production of the antibiotic and saving thousands of wounded soldiers from deadly infections.

THE BIONIC FUTURE

INVENTING THE FUTURE

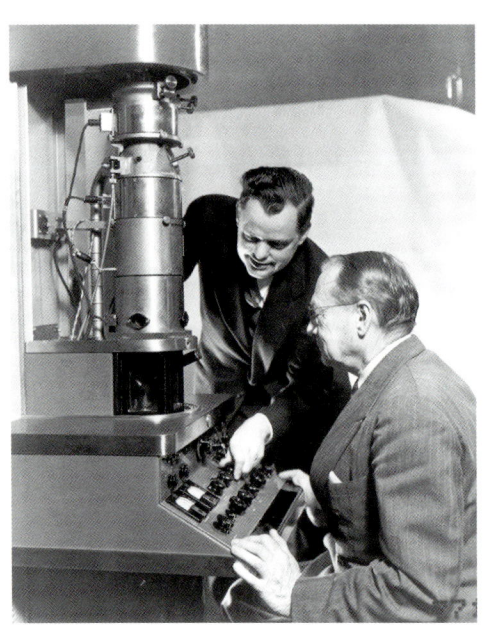

ABOVE: James Hillier, Richard Snyder, and Vladimir Zworykin operate the world's first scanning electron microscope during the war, with images generated on the Labs' facsimile printer.

LEFT: Hillier shows the controls of an early transmission electron microscope to Vladimir Zworykin.

RIGHT: World's first disposable hearing aid, developed by Sarnoff and introduced by Songbird Hearing in 2001.

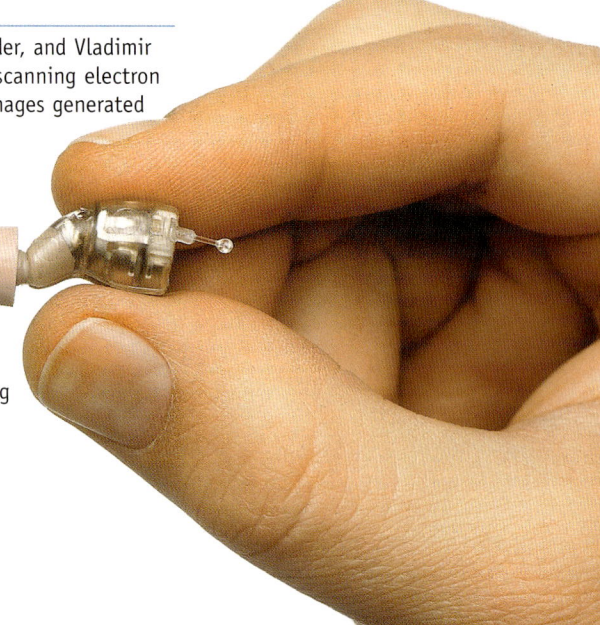

THE BIONIC FUTURE

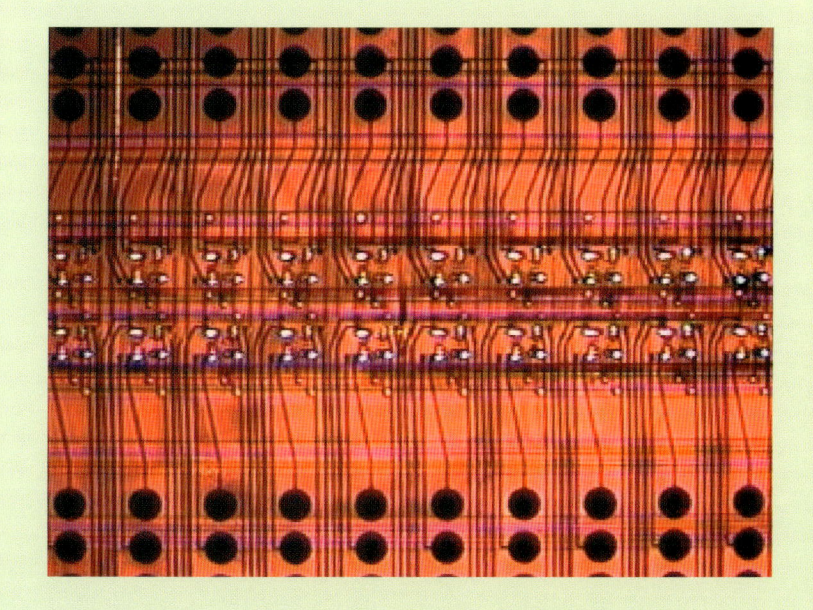

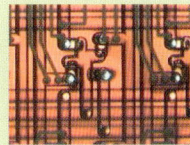

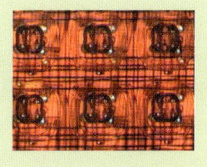

LEFT AND ABOVE: Magnified views of the pumps, vias, and wells on a microfluidic chip (see photo bottom left corner).

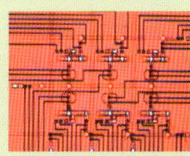

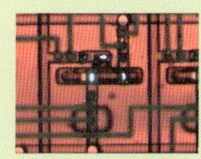

Above: *Left*: Early six-cell microfluidics chip from 1996, then the most complex such structure in the world. Electrostatic pumps moved chemicals and reagents across the chip to oval cells to conduct combinatorial chemistry reactions. *Right*: Close-up of a reaction cell from this chip.

BELOW: Microfluidic chips, developed by Sarnoff in the late 1990s, put a combinatorial chemistry lab on a device the size of a business card. Electrostatic pumps with no moving parts sent samples and reagents to test sites on the chips. The results were recorded by lasers.

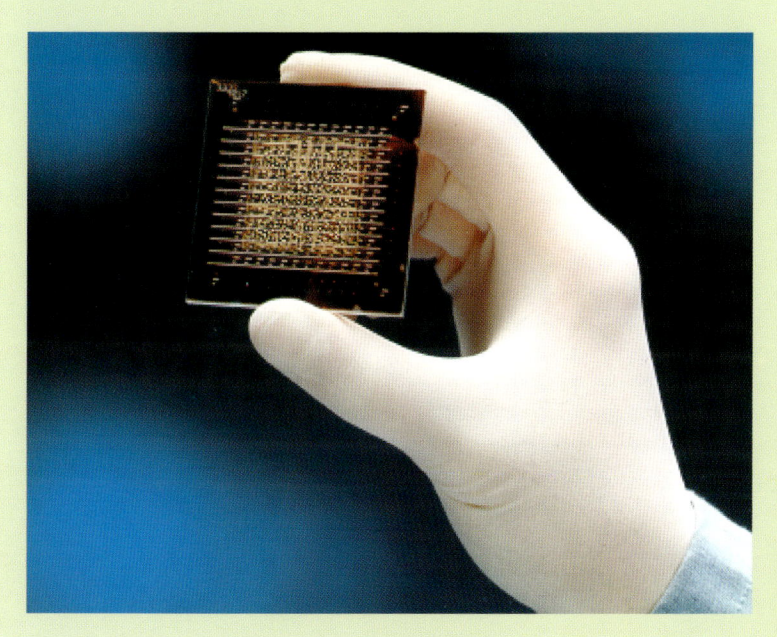

BELOW: A 24-cell microfluidics chip for chemical synthesis, created in 1999, showing reservoirs for samples and reagents.

INVENTING THE FUTURE

Extending the Power of Sight

ELECTRONIC CAMERAS AND IMAGERS

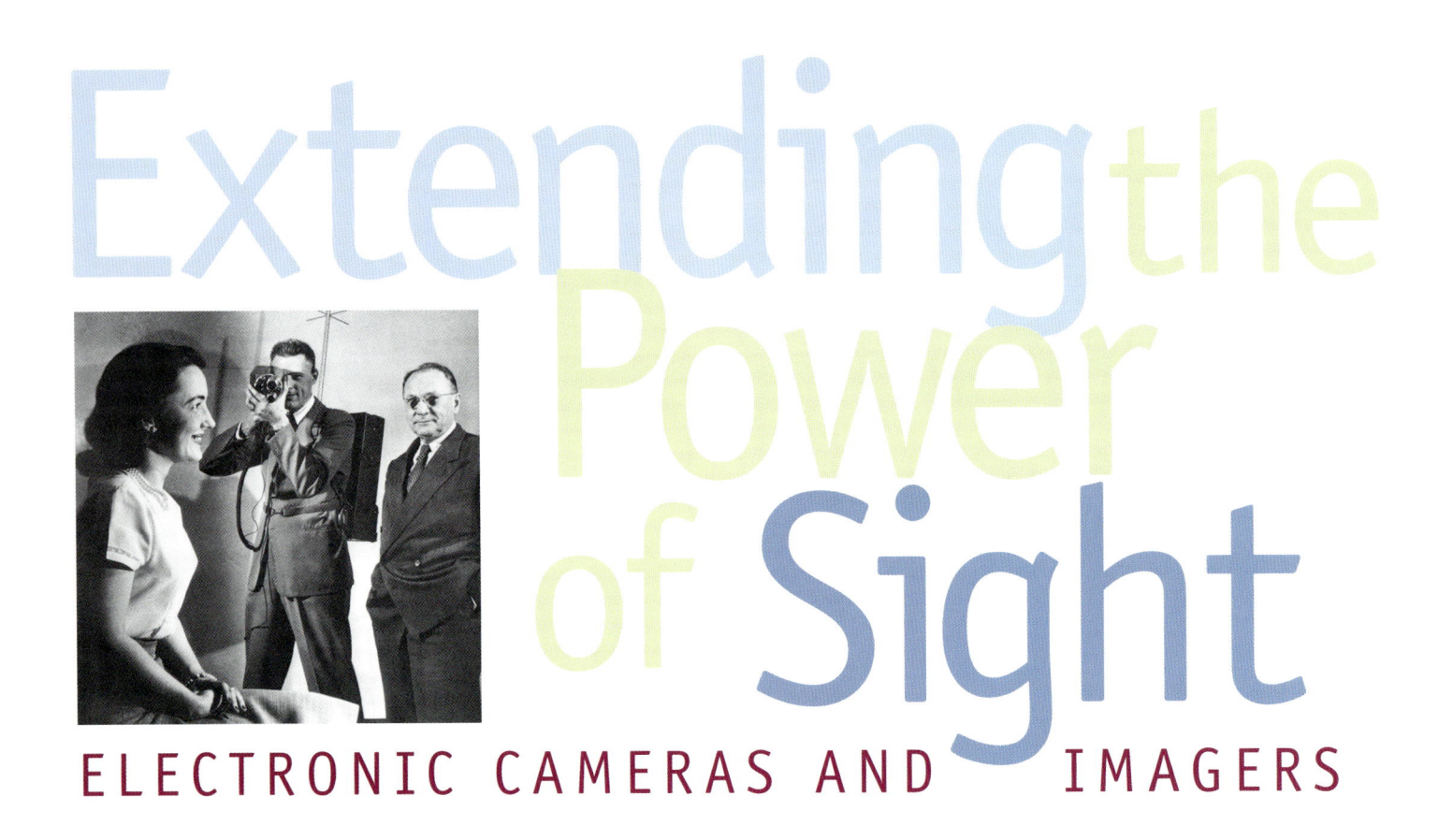

Electronic cameras are the greatest advance in human perception ever achieved.

Painting and film show us what someone else saw at a moment in the past. With electronic cameras we see things as they exist now, and events as they unfold, even over vast distances.

The implications are immense. Today cameras let us watch history being made, live and unedited. These electronic eyes let us observe military actions from a safe distance, or robots at work in a hostile environment. Because we see these activities we can directly influence their progress and results.

The idea of turning electronic cameras into the eyes of a control system came early. In the 1930s RCA proposed using TV cameras in military guidance systems to defense officials. In 1943 one of the first projects at today's Sarnoff Corporation was developing the Image Orthicon camera pickup tube for military use. Pictures from "Immy" tubes in glide bombs and B-17 drones guided attacks on the enemy in World War II.

The Immy inaugurated Sarnoff's leadership position in imaging technology. Over the past 60 years Sarnoff innovations have included the development of the first single-sensor color camera, the X-Y scanning approach used in CMOS imagers, and the standard output stage for charge-coupled devices (CCDs). Sarnoff has also designed and manufactured multispectral imaging chips and the world's first MEMS-based infrared (IR) sensor.

Sarnoff camera technology has been used in such applications as spectroscopy, astronomy, space imaging, and ultraviolet (UV) inspection equipment for semiconductors.

ABOVE: Vladimir Zworykin oversees a 1951 demonstration of RCA Labs' "Walkie-Lookie" TV camera and backpack transmitter. Newscasters used the system at the 1952 and 1956 presidential conventions.

SIGNIFICANT SARNOFF IMAGER AND CAMERA INNOVATIONS

1943 Image orthicon ("Immy") camera pickup tube, 100 to 1000 times more sensitive than previous designs, enabling use with moderate light; the Emmy® award was named for it.

1950 Vidicon tube, a compact pickup tube for lower bandwidth applications, the standard for industrial cameras into the 1980s.

1952 Tricolor vidicon, the first single-sensor color camera; its innovative striped color filter to derive RGB signals from a single device is still the standard approach

1966 Thin-film transistor (TFT) battery-operated camera, introducing the X-Y scanning architecture used in today's CMOS imagers

1971 Invention of floating diffusion output stage for CCDs to improve low-light sensitivity, now the standard circuit in camcorders and digital still cameras.

1976 Adaptation of CCD sensor to TV comb filter use.

1984 First commercially available all-solid-state broadcast camera, the CCD-1, which used CCDs to exceed the performance of the best tube cameras

1990 World's first solid-state spectrometer sensor, a combination of CMOS and spectroscopic CCD processes; delivers much higher system performance.

1993 Highest resolution (640 x 480 pixel), highest sensitivity IR imager and camera then available, based on platinum/silicon (PtSi) technology.

1993 First high-performance megapixel back-illuminated imager, featuring low noise and high frame rates.

1997 CMOS active pixel sensor (APS) imager with high 110 db dynamic range for low-cost "camera-on-a-chip" products.

2001 1024 x 1024 deep ultraviolet (UV) CCD with high quantum efficiency (sensitivity) and high frame rate for inspection of masks for deep submicron ICs.

2002 CCD-based deep UV camera module operating at up to 1000 frames/second.

2003 World's first full-resolution (320 x 240 pixel) uncooled IR imager based on microelectromechanical system (MEMS) technology.

2003 Array of 112 Sarnoff imagers in Mt. Palomar QUEST telescope show the most distant object yet discovered in the solar system.

GLOBAL IMPACT

The availability of high-quality, low-cost electronic imagers and cameras has generated an explosion of imaging-based products.

- Electronic imagers have created a video and digital photography market that generates hundreds of billions of dollars worldwide.
- Cell phone cameras have created a new convergence product, letting millions of users instantly transmit visuals along with their voices in business and personal communication.
- Early "see-in-the-dark" IR cameras on luxury cars are the first wave of imaging-based automotive convenience and safety features.
- Surveillance cameras multiply the effectiveness of security operations, protecting people from criminal activities and helping to identify wrongdoers.
- Tiny cameras inside the body guide doctors through minimally invasive surgery, reducing trauma and recovery time for patients.
- IR sensors power medical thermometers, industrial heat sensors, and thermal imagers for safety and security use and in firefighting applications.
- Machine vision systems use cameras for product identification, measurement, inspection, and robot guidance.

Electronic cameras let us view phenomena that are too fast, too slow, too far away, too small, or in wavelengths too short or long for the human eye. They give robots eyes. Some day they may restore sight to the blind by linking directly into the human nervous system. The potential is as far-reaching as the power of sight itself. Sarnoff continues its groundbreaking work in developing new, more powerful, more useful imaging devices.

RELATED TOPICS

Television and Video: see *The Video Revolution; The World on Display*
Machine vision: see *Teaching Computers to See*
Semiconductor imagers: see *The Microelectronics Miracle*

ABOVE: Active pixel sensor CMOS imager, c2001.

INVENTING THE FUTURE

EXTENDING THE POWER OF SIGHT

ABOVE: View of a street scene taken with infrared light. RCA Labs pioneered electronic IR imagery in the 1940s.

LEFT: Albert Rose takes the measure of the image orthicon, which he designed to incorporate the best features of earlier video camera tubes. He, Paul Weimer, and Harold Law made it small enough for use in guided missiles during World War II.

INVENTING THE FUTURE

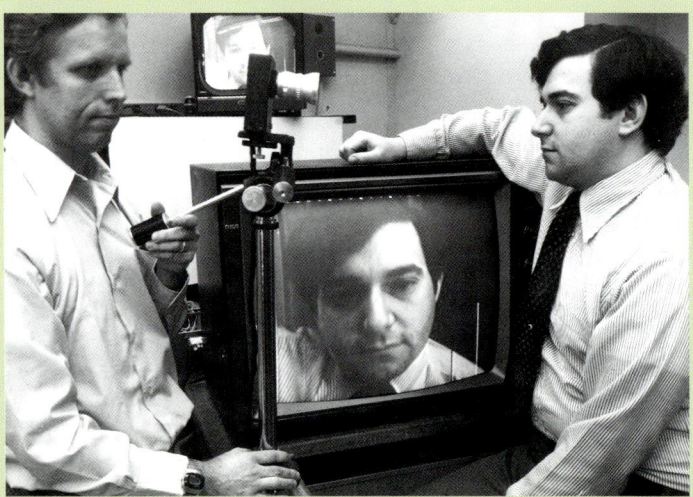

TOP LEFT: A demonstration of a CCD camera in 1975 recalls the use of Felix the Cat in RCA's first TV camera tests in 1928.

ABOVE: The image orthicon tubes in RCA's television cameras merited a cover story in RCA's Broadcast News magazine for Spring, 1947.

LEFT: Peter Levine is on candid CCD camera in 1977.

RIGHT: Blinc camera used Sarnoff's active pixel CMOS sensor to achieve 110 db dynamic range, revealing details in shadows and highlights simultaneously.

EXTENDING THE POWER OF SIGHT

INVENTING THE FUTURE

EXTENDING THE POWER OF SIGHT

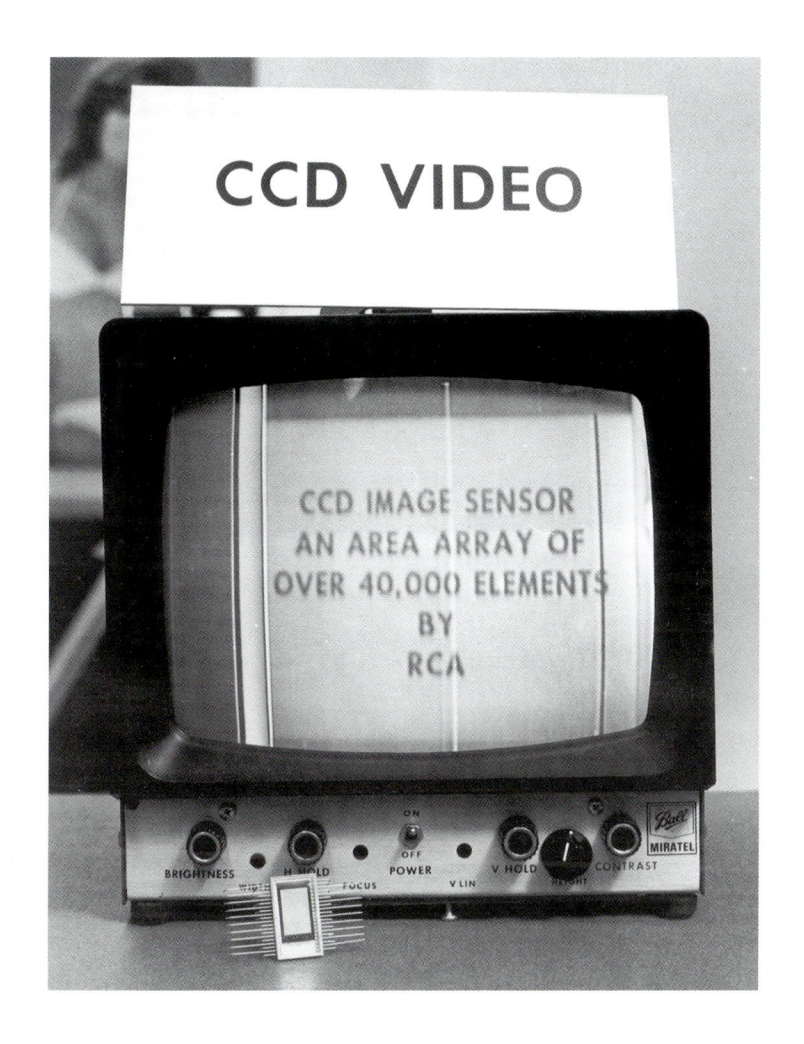

TOP LEFT: RCA announces a CCD-based image sensor five years after the TFT camera pictured at left.

TOP RIGHT: Winthrop Pike focuses the TFT camera for which he devised the circuitry, 1967.

RIGHT: The one-megapixel CAM1M100 deep-UV to near-IR camera, announced in 2005, uses a Sarnoff CCD imager for 100 frame/second high-throughput inspection and biomedical applications.

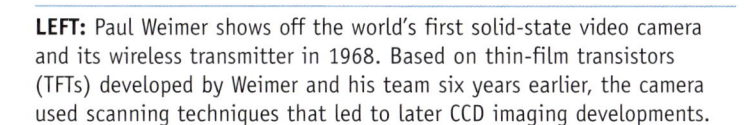

LEFT: Paul Weimer shows off the world's first solid-state video camera and its wireless transmitter in 1968. Based on thin-film transistors (TFTs) developed by Weimer and his team six years earlier, the camera used scanning techniques that led to later CCD imaging developments.

INVENTING THE FUTURE

ABOVE: George Morton leads a review of infrared imager projects during World War II.

RIGHT: An array of 112 highly sensitive Sarnoff CCD imagers, installed in the QUEST astronomical telescope at the Palomar Observatory in 2003, helps scientists find quasars, asteroids, and supernovas in the depths of space.

EXTENDING THE POWER OF SIGHT

Larry Boyer holds a transistorized Vidicon-based 1956 minicam combined with a backpack transmitter, which freed news reporters to cover a story no matter where it took place.

INVENTING THE FUTURE

Transmitting Intelligence
COMMUNICATIONS AND COMPUTING

This is an interconnected age.

It began in 1912, when "wireless" meant maritime Morse code, and its use to save lives during the Titanic disaster awoke the world to the importance of electronic communications. A Marconi telegrapher named David Sarnoff watched the drama unfold, and played his own small role.

By 1921 wireless had become the "radio music box," as Sarnoff called it in a famous 1915 memo proposing the concept. Like magic, it brought voices and music – news, events, and entertainment – through the ether to faraway towns across America.

In 1922 Sarnoff was talking about "the transmission of intelligence" as the wave of the future. In 1964 he predicted, "Through communications satellites, laser beams, and ultraminiaturization, it will be possible by the end of the century to communicate with anyone, anywhere, at any time, by voice, sight, or written message." He was right. No matter where we are, computer and communications technology keeps us in the know, in touch, and in control. It gathers, processes, and distributes intelligence.

Sarnoff researchers have played a seminal role in creating this new age. They developed basic communications devices, advanced radar systems, satellite equipment, computer memory, programs, microprocessors, and wireless information networks. Their work led to innovations in cell phones, satellite TV, microwave links, and HDTV. The Sarnoff legacy of technology continues to power the interconnected Earth.

ABOVE: Jan Rajchman and an associate examine one of the first 10-kilobit magnetic core memories in 1956. Rajchman co-invented the technology, used into the 1970s as random-access memory for computers.

SIGNIFICANT SARNOFF COMMUNICATIONS AND COMPUTING INNOVATIONS

1942 TV-based SHORAN (Short Range Navigation) system, precursor of modern mapping tools, accurate to a foot in 60 miles, also used for bomb guidance.

1942 Noise factor concept and method to measure sensitivity in radio receivers.

1944 Narrow beamwidth radar antenna to eliminate interference from other signals.

1945 Traveling wave tube, high-power, low-noise wideband amplifier for microwave communications and radar, still in use today.

1947 Selectron vacuum tube 256-bit computer memory, conceptual forerunner of RAM chips.

1950 Typhoon analog computer for naval missile fire control.

1952 Ferrite core matrix memory for mainframe computers.

1959 Commercial production of tunnel diodes.

1961 RF-capable transistors for radio use.

1971 Computer-aided RF circuit design program, forerunner of CAD software.

1973 Microwave GaAs power FET, solid state device for amplifying C-band communications.

1978 Testing and modeling of microwave circuits using the Curtice large-signal model, still in use today.

1982 GaAs FET Microwave Amplifiers, the first solid-state power amplifiers (SSPAs) to be used in communications satellites in space.

1990 Compact SSPA for the TIROS weather satellite and the Space Station Freedom.

1984 Silicon millimeter wave devices to extend communications into mm wavelengths.

1985 Princeton Engine parallel processing supercomputer, later the basis of a digital video server.

1992 Developed direct digital satellite TV standard and built prototype system; became fastest growing consumer product in history.

1992 Radio frequency identification (RFID) tag, a passive credit card-sized device to transmit and receive information at up to 20 feet.

1995 Locomotive speed sensor, highly accurate Doppler radar/microwave device to improve hauling efficiency.

1998 Cyclone cluster supercomputer, built from networked generic processors, now used in protein modeling for pharmaceutical research.

2001 Ad Hoc networking protocols for creating instant self-configuring wireless networks of sensors, computers, and other devices.

2003 GPS-based system for bringing new electrical substations on line quickly and safely.

2005 SVD-1000 and 2000 stray voltage detectors for identifying potentially dangerous electric current in street-level objects.

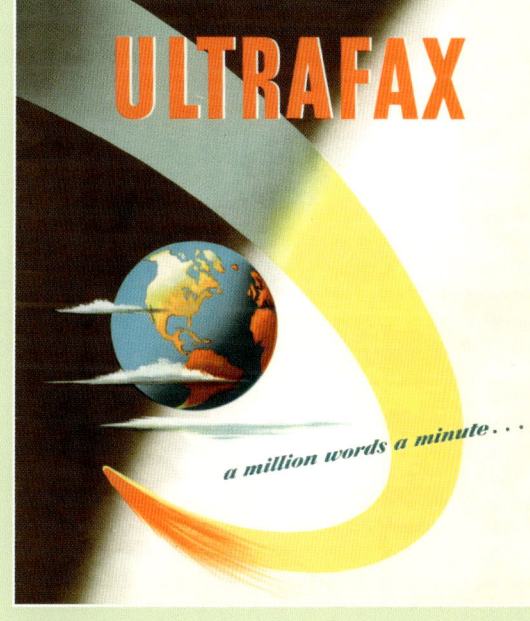

GLOBAL IMPACT

Computers and communications are the basis of the global economy. Companies use computers to design the products we buy. They use communications to gather market data from around the world, and to link their design and manufacturing teams across continents. Military and commercial aircraft use radio for communication, radar to determine location and range, and computers to plot courses and run onboard systems.

The same technologies have helped improve the lives of billions of individuals. Consumers watch news on satellite TV, get directions from global positioning satellites (GPS) in their cars, rely on the microwave and antenna technologies that power the cell phone system, and connect to the Internet through wireless routers.

The commercial impact of "the transmission of intelligence" is virtually incalculable. Its transformation of human life for the better is undeniable.

RELATED TOPICS

Lasers for telecommunications: see *Information on a Beam of Light*
Microprocessor technology: see *The Microelectronics Miracle*
Computers in medicine: see *The Bionic Future*

ABOVE: RCA researchers had been developing wireless facsimile since the 1920s when they demonstrated Ultrafax, a 1948 system that sent the full text of *Gone with the Wind* from New York City to the Library of Congress in two minutes.

INVENTING THE FUTURE

TRANSMITTING INTELLIGENCE

ABOVE: Max Mesner finishes inspection in 1960 of RCA's pioneering TIROS weather satellite, a project that began at the Princeton Labs before it evolved into the AstroElectronics Division.

LEFT: Sister Mary Hilda processes new Medicare patients on the RCA 301 computer at Mount Carmel Mercy Hospital in Detroit. RCA's computer business developed out of computers built at the Labs for the U. S. military in the 1940s.

RIGHT: *Shown at actual size:* An RFID microtransponder the size of a grain of salt, developed in 1998, contains an IC with photocell, radio transmitter, and logic circuitry. When illuminated by a laser to provide power, it transmits a unique identifying sequence.

INVENTING THE FUTURE

TOP LEFT: This microwave frequency multiplier for the Apollo lunar landing expeditions was a product of the Microwave Advanced Research Laboratory at the David Sarnoff Research Center.

ABOVE: Robert Matturo holds up one of the processor boards for the Princeton Engine in 1989, Sarnoff's massively parallel computer for real-time modeling of broadcast and high-definition video systems. It became the basis for the company's pioneering cable TV video-on-demand system.

LEFT: After World War II, RCA Laboratories designed, built, and tested the Typhoon computer for the U. S. Navy's guided missile program as a complement to M.I.T.'s Whirlwind computer for the U. S. Air Force.

RIGHT: Demonstration of the Labs' Homefax information system, 1968, in which news, weather reports, recipes, and other information could be delivered over the air to selected households.

TRANSMITTING INTELLIGENCE

INVENTING THE FUTURE

ABOVE: An RCA Relay satellite undergoing testing. Relay, a product of RCA AstroElectronics Division with consulting assistance from the Labs, transmitted the first color television programming by satellite in 1962.

ABOVE: Sarnoff's 1960 electronic highway test track in Princeton showed how an automobile, provided by General Motors Research Laboratory, could interact with circuits implanted in the roadbed to control steering and prevent collisions, even with unequipped vehicles.

BELOW: Sarnoff's SVD-2000 stray voltage detector uses a sophisticated sensor array with proprietary electronics and software to detect dangerous voltages in manhole covers, utility boxes and lamp poles that could pose a hazard to passersby.

ABOVE: Nate Goldberg plots the courses of a model bomber and missile to demonstrate what the Typhoon computer could do electronically in 1950.

BELOW: Inside the truck a specially-equipped computer displays voltage readouts from lamp poles and other objects, along with video of the scene, and alerts the operator to potential stray voltage hazards.

INVENTING THE FUTURE

ABOVE TOP: Sarnoff-designed Cyclone cluster computer with high-resolution tiled display, c1994

ABOVE BOTTOM: Lew Stetz and Joe Zenel show off the Labs' new 5-meter antenna receiver in 1977.

LEFT: Krishna Jonnalagadda admires one of Sarnoff's C-band satellite communications dishes in 1990, a tool that helped lay the groundwork for digital TV broadcasting by satellite.

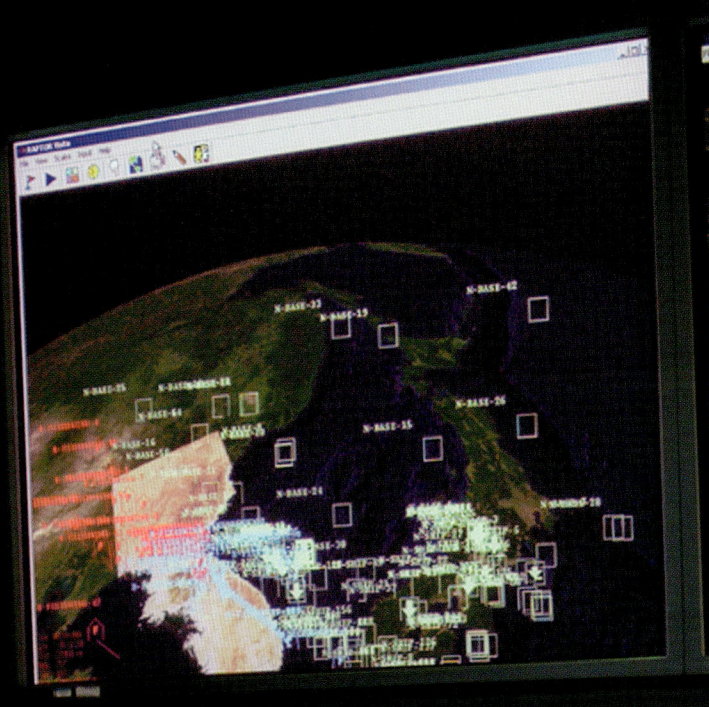

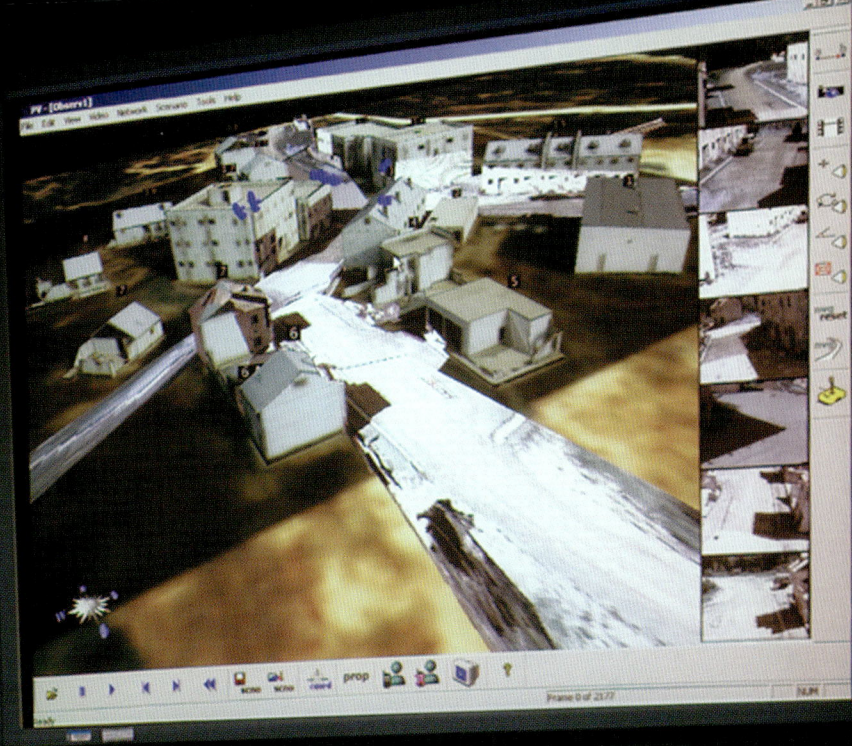

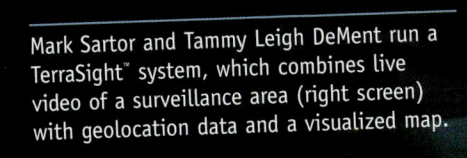

Mark Sartor and Tammy Leigh DeMent run a TerraSight™ system, which combines live video of a surveillance area (right screen) with geolocation data and a visualized map.

INVENTING THE FUTURE

Elements of Innovation

ELECTRONIC MATERIALS AND PROCESSES

Stone Age, Bronze Age, Iron Age, Steam Age… Electronics Age.

We live in an era of human progress that surpasses any of our previous great leaps forward.

Like earlier periods of advancement, the Electronics Age uses newly discovered materials to create innovative technology and products. RCA researchers and their Sarnoff successors have made crucial contributions right from the start.

The list of Sarnoff-developed materials could fill a book. It includes phosphors, liquid crystals, amorphous silicon, semiconductors, superconductors, and organic (plastic) electronics. They are the foundation of modern electronics. Sarnoff also developed many of the processes to manufacture and use these materials.

It started in the 1930s, when an RCA phosphor gave cathode ray tubes (CRTs) and fluorescent bulbs the ability to produce a more natural white light. Since then Sarnoff has produced phosphors for the first color CRT and all that followed; for plasma, electroluminescent (EL), and field emission displays (FEDs); and for anti-counterfeiting and product identification use.

Sarnoff researchers developed virtually all the III-V semiconductor compounds used to make solid-state electronics. They also pioneered the process that made reliable LEDs possible, and developed amorphous silicon for solar cells and LCD drive transistors.

The 1990s saw the development of ceramic/metal composites for high-power, high frequency electronic devices, and of microfluidic silicon structures for chemical analysis on a chip.

Sarnoff researchers today are developing organic transistors for low-cost, flexible organic LED displays, and phosphors for LED-based electronic lighting.

ABOVE: RCA Labs collaborated with RCA Victor's Advanced Development team to build some of the world's first transistor radios in 1953.

IMPORTANT SARNOFF MATERIALS AND PROCESS INNOVATIONS

1942-1944 Phosphors developed for first CRT radar screens, crucial to war effort.
1950 Completed RGB phosphor set for world's first color TV CRT.
1953 Magnetic tape for professional recording of monochrome and color video.
1966 RCA red, a rare-earth phosphor that became the standard for CRTs, and the process to manufacture it.
1966 Green phosphors for early electroluminescent displays.
1970 Standard Clean 1 and Standard Clean 2 (SC1 and SC2) wafer cleaning solutions, known as RCA Cleans and still in wide use.
1962 Metal oxide semiconductor field effect transistor(MOSFET), the basis of C (complementary) MOS process.
1962 Discovery of electro-optical properties of nematic crystals, the basis for LCD technology.
1962 Superconducting niobium-tin magnetic materials and process for high-power solenoids; delivered both research and commercial units.
1964 CMOS process, used today to manufacture most ICs and DRAM chips.
1965 Nematic liquid crystals doped with pleochroic color dyes, used in dynamic scattering liquid crystal information display, the first commercial LCD.
1966 Silicon-germanium (Si-Ge) alloy for thermoelectric power generation, used in power source for such deep space missions as Lincoln, Voyager I and II, Galileo, Ulysses, and Cassini.
1977 Silicon-on-sapphire process, used to make SoS ICs for space missions.
1977 Amorphous silicon solar cell patented.
1982 Amorphous silicon solar cell achieves 10% light-to-electricity efficiency.
1989 Polysilicon active-matrix liquid-crystal display (AMLCD).
1992 First amorphous silicon thin-film transistor fast enough to drive LCDs; matrices of a-SI:H TFTs are now standard in laptop PC displays.
1993 Blue EL phosphor, last color element needed for a full-color EL display.
1994 Ceramic on Metal, world's first low cost, low expansion, high thermal conductivity laminate for direct mounting of high power Si and GaAs bare dies for multichip modules (MCMs).
1994 Phosphors to uniquely identify products or documents for anti-counterfeiting and product tracking.
2001 Co-development of a video-capable, flexible plastic LCD display printable by ink-jet, based on an organic TFT.

GLOBAL IMPACT

Materials are the invisible substrates of commerce. We see the ultimate products, not the stuff of which they're made.

That makes it difficult, if not impossible, to assess the socioeconomic impact of new materials. When they have multiple applications, as many do, the difficulties are compounded.

Other sections of this book talk of the products these materials made possible, and show how those products helped create new industries.

Total sales and other quantitative and qualitative outcomes indicate the social and economic benefits that new products deliver. So perhaps the easiest way to indicate the impact of Sarnoff's innovations in materials is to list product areas that would not exist without them. They include:
- Television, especially color TV.
- Computer monitors, including CRTs and LEDs.
- Modern electronics, including radios, computers, cell phones, anything that uses transistors, ICs, RAM, microprocessors.
- Silicon solar cells.

There are many more, but the point is clear: Sarnoff's innovations in materials have created a substantial portion of the bedrock technology on which we have built our electronic age.

RELATED TOPICS

CRTs, LCDs: see *The Video Revolution; The World on Display*
Transistors and ICs: see *The Microelectronics Miracle*
Lasers: see *Information on a Beam of Light*

ABOVE: Robert Simms inspects the electrostatic deposit CRT phosphors on the face plate, a process commercialized by one of Sarnoff's clients.

INVENTING THE FUTURE

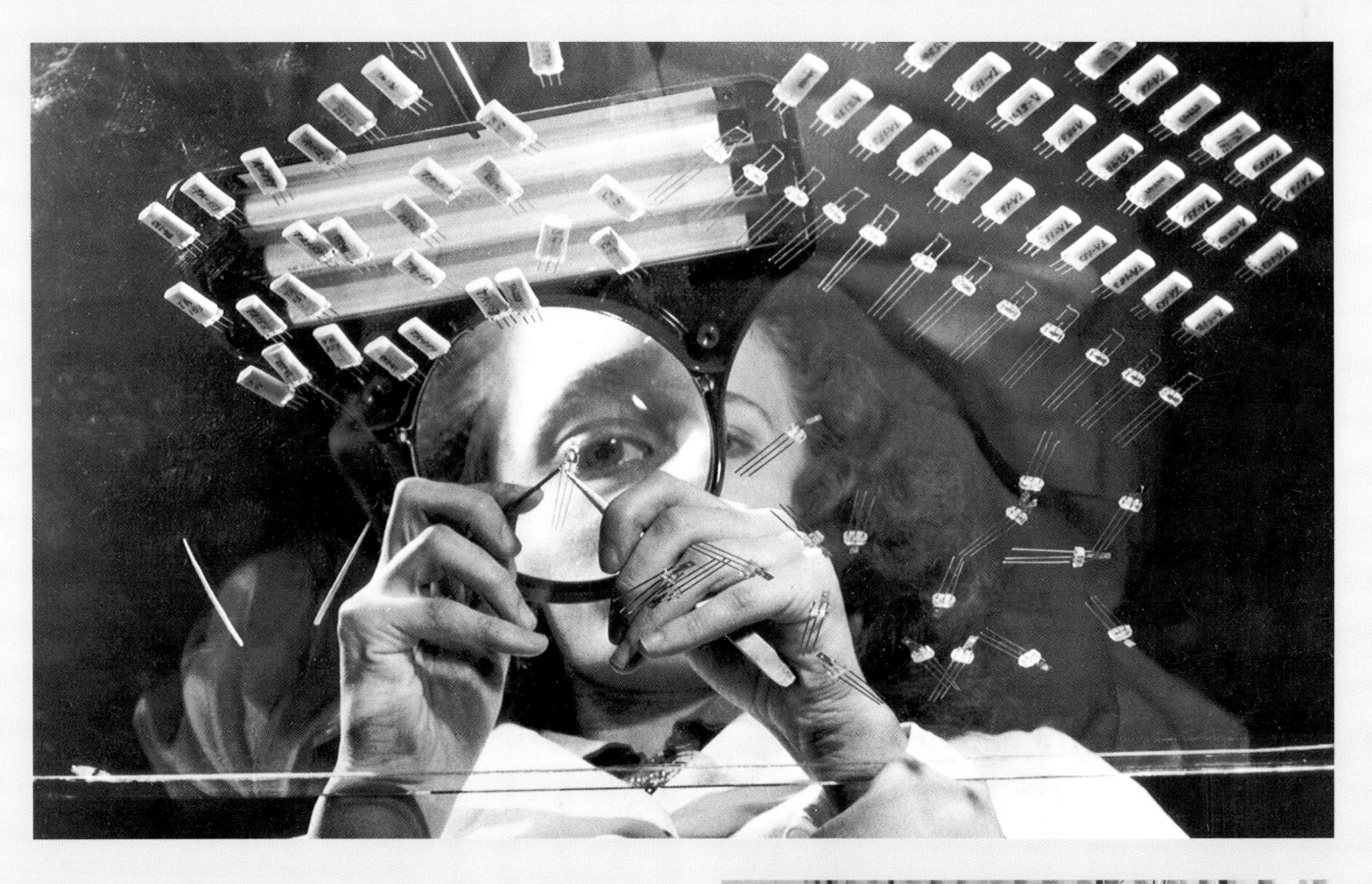

ABOVE: Assembling alloy-junction transistors at the Harrison, NJ RCA factory in 1952. Labs staff developed the techniques used there to make materials and components.

RIGHT: Schuyler Christian oversees the pulling of crystal germanium for transistors, 1952.

ELEMENTS OF INNOVATION

ABOVE: Jerome Kurshan built RCA's first transistor within a month of Bell Telephone Laboratories' announcement of the device in June 1948.

RIGHT: A technician loads germanium alloy-junction transistors designed at the Labs into an oven at RCA's pilot plant.

BELOW LEFT: Bob Hesser and Frank Kolondra adjust the controls on an ion implanter in the late 1970s, used for precision doping of semiconductor wafers with other elements.

BELOW RIGHT: Demonstrating the Labs' transistorized wireless keyboard in 1952.

INVENTING THE FUTURE

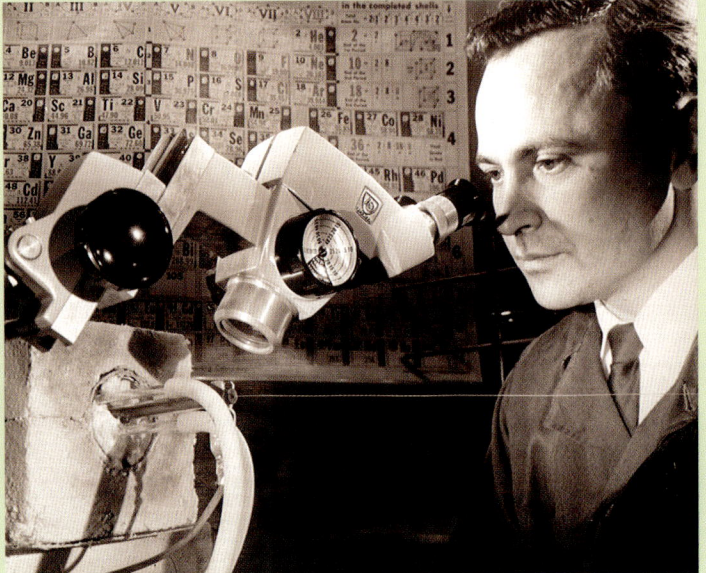

ABOVE: David Carlson oversees the deposition of amorphous silicon on a solar cell in 1977. Starting in 1973, he led the efforts that increased the photoelectric conversion efficiency from 1% to 10%.

LEFT: Joseph Hanak views some of the niobium-tin superconducting wire that he and his colleagues developed and transferred to RCA Harrison for commercial sale in 1960. RCA led the world in this field throughout the 1960s based on research and early development at the Labs.

RIGHT PAGE (CLOCKWISE):
1) David Sarnoff demonstrated this atomic battery, made of a semiconducting wafer and thin coat of strontium-90 on the cylinder, in 1954.
2) Lighting a Christmas tree with power from amorphous silicon solar cells during the energy crisis of 1979.
3) Joseph Abeles in 1963 with a fan powered by his team's silicon-germanium alloy, which converts heat to electricity. RCA Harrison and Lancaster made and sold it for use in deep-space satellites such as Voyager and Galileo.
4) Jacques Pankove, Charles Mueller, and Loy Barton show off their high-frequency alloy junction transistors and their application to radios in 1952.

ELEMENTS OF INNOVATION

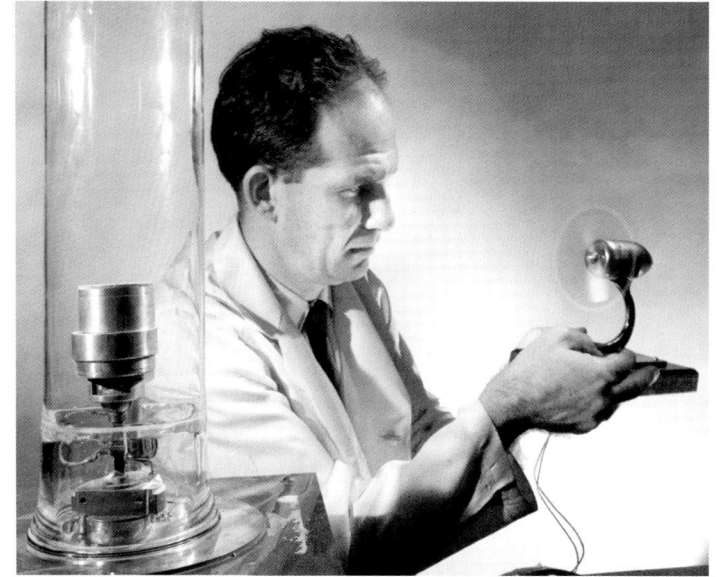

INVENTING THE FUTURE

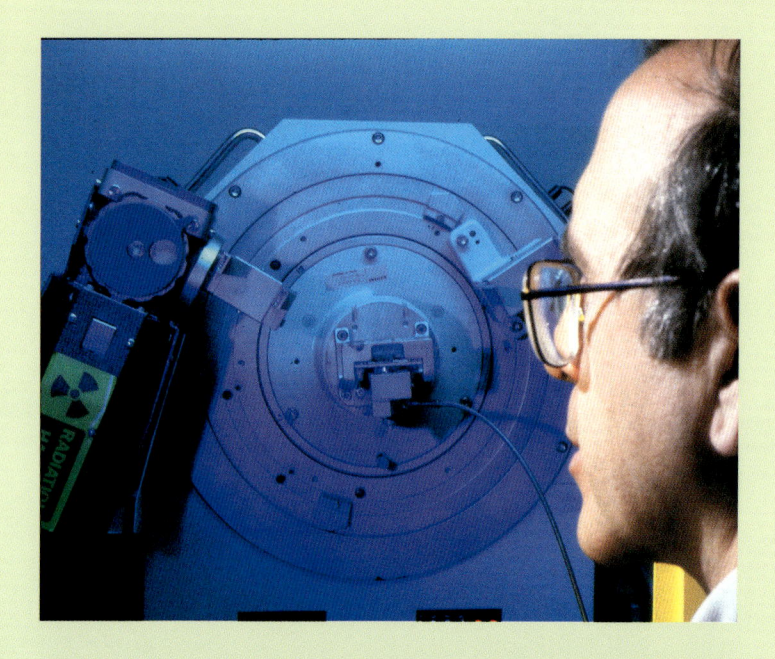

RIGHT: Prepping wafers in one of Sarnoff's clean rooms, 2003.

LEFT: S. L. "Jock" McFarlane uses an x-ray diffractometer in the Materials Characterization Group to characterize epitaxial layers in 1988.

BELOW LEFT: Silicon-on-sapphire chips on a wafer at Sarnoff, 2004.

BELOW RIGHT: Hollow-cathode plasma deposition of thin-film diamonds, 1988.

ELEMENTS OF INNOVATION

INVENTING THE FUTURE

ELEMENTS OF INNOVATION

ABOVE LEFT: Lori Hewitt of Sarnoff's Materials Characterization Group operates an ion accelerator used in Rutherford Backscattering Spectrometry in 1988.

ABOVE RIGHT: Mike Grote measures the phosphor output of a CRT, c1993.

RIGHT: A Sarnoff technician with some of the highly specialized phosphors the Labs created for displays, product identification, and other applications.

LEFT: Randy McCoy is reflected in the window of the Display Materials Research Lab in 1988. The lab produced a steady stream of innovations in color displays.

INVENTING THE FUTURE

Jim DeMarco holds a sample of Sarnoff's low-temperature co-fired ceramic on metal (LTCCM) backplane for electronic assembly. LTCCM handles very high frequencies and high heat, and can include embedded passive components.

ELEMENTS OF INNOVATION

The David Sarnoff Research Center, Princeton, NJ. Home of RCA Labs since 1942, and headquarters of Sarnoff Corporation since 1987.

Stay Tuned

THE FUTURE IS STILL AHEAD

This book is a rare look back. The people who make up Sarnoff's most valuable resource are by nature more interested in the prospects and possibilities for the next big innovation than they are in dwelling on what's already been achieved.

But they are proud of their company's history, and like all true innovators they know that the seminal discoveries of yesterday are preludes to the breakthroughs of tomorrow.

Perhaps it is impossible to predict what those breakthroughs are, or when they will come. We can only look at work in progress and project its results.

Today optoelectronics and vision technology are two of the most promising areas of development, and Sarnoff researchers are focusing their attention there. But they are not neglecting other spheres of innovation. As this book goes to press, they are pushing forward on innovations amazing as the ones listed in previous chapters, including:

- Faster, more capable imagers and cameras for scientific and industrial applications.

- Advanced security and surveillance systems that integrate video with other data and automatically format it for PDAs and PCs, so all members of a team can share it in real time.

- Vital signs monitors that track ambulatory patients in a hospital or at home.

- More effective electrostatic discharge protection for ICs and other chips.

One thing has not changed. For over 60 years clients have relied on Sarnoff Corporation to develop not just new technology, but marketable innovations for successful new products. Sarnoff remains dedicated to this goal.

INVENTING THE FUTURE